I0065253

DIGITAL GOVERNANCE

A Practical Guide for CEOs of Large Enterprises

Ahmad Almulla
and
Arun Tewary

PASSIONPRENEUR®
P U B L I S H I N G

Digital Governance
Copyright © 2022 Ahmad Almulla and Arun Tewary
First published in 2022

Print: 978-1-76124-046-1
E-book: 978-1-76124-047-8
Hardback: 978-1-76124-048-5

All rights reserved. No part of this book may be reproduced, stored in a retrieval system, or transmitted by any means (electronic, mechanical, photocopying, recording, or otherwise) without written permission from the author.

Because of the dynamic nature of the Internet, any web addresses or links contained in this book may have changed since publication and may no longer be valid. The information in this book is based on the author's experiences and opinions. The views expressed in this book are solely those of the author and do not necessarily reflect the views of the publisher; the publisher hereby disclaims any responsibility for them.

The author of this book does not dispense any form of medical, legal, financial, or technical advice either directly or indirectly. The intent of the author is solely to provide information of a general nature to help you in your quest for personal development and growth. In the event you use any of the information in this book, the author and the publisher assume no responsibility for your actions. If any form of expert assistance is required, the services of a competent professional should be sought.

Publishing information
Publishing, design, and production facilitated by Passionpreneur Publishing, A division of Passionpreneur Organization Pty Ltd, ABN: 48640637529

www.PassionpreneurPublishing.com
Melbourne, VIC | Australia

I would like to dedicate this book to my late father who believed in me and made me believe in myself, set a vision for me, inspired me with words of wisdom, and guided me all the time. Remembering his words still inspires me.

—Ahmad

I would like to dedicate this book to my parents and my younger brother who are in Heaven, my wife Premshila, my daughters Pratiksha and Nandita, my son Gaurav and my three sisters, Madhuri, Maya and Mamta.

—Arun

TABLE OF CONTENTS

Introduction to Digital Governance   1

Strategy and Execution   21

The Supporting Ecosystem   41

Change Management   53

Frameworks, Standards, and the 3 Ps   61

Performance Management   77

Cyber Security   95

Emerging Technologies   113

Talent Management   123

Data Governance   133

Conclusion   145

Glossary   151

References   155

Acknowledgments – Ahmad   161

Acknowledgments – Arun   163

Extras   165

About The Authors   167

INTRODUCTION TO DIGITAL GOVERNANCE

Successful digital transformation through effective and practical governance model.

Digital Transformation and Digital Governance

Would you live in a country where there is no government? Certainly not. In the absence of a government, who ensures that laws and regulations are enforced? Without a government, there could be anarchy. The term "governance" refers to control, rule, and power. In the context of digital transformation, the same is true. To ensure that every digital transformation journey is successful, a solid governance system must exist so that objectives are met, investments are protected, and projects are completed on time, within budget, covering the full scope, and meeting the expected quality. Without governance, results cannot be predicted; similarly, with a governance model that is too complex, results cannot be predicted.

What is needed is a governance model that is practical, comprehensive, simple to use, and ensures that goals and objectives are met leading to successful transformation. Such a model is called "Digital Governance". The need for such a model comes with the trend of increased shift towards the adoption of digital technologies which require large amounts of investments, and with the existence of a fast-changing technology landscape.

The best place to start is by defining the foundation of digital governance concepts and fundamentals in a simple and practical manner.

The world is passing through massive and unprecedented digitalization. The pace is so rapid that many times it becomes incomprehensible and, hence, complex. Digitizing, digitalization and digital transformation are different terms with totally different meanings, interpretations, and relevance. Unfortunately, those terms are not well understood and are commonly misused.

Digitizing is the process of converting information from a physical format into a digital one.

Digitization is the use of digital technologies and data to improve existing business processes and create a digital culture where digital data is at the core of the process.

Digital transformation is a journey towards digital business. It involves a redefinition of the business operating model. Digital transformation includes both digitization and digitalization, using these two processes to transform the way that an enterprise conducts its business. In other words: digital transformation leads to a complete evolution of the business model.

Digital Transformation Today

Digital transformation is on the agenda of the board of every business across all industries. Many businesses have started this journey, many are about to start, and some have advanced quite a bit. This trend started a few years ago but accelerated dramatically upon the emergence of COVID-19. Many studies have been conducted that support this trend.

One study showed that the pandemic has accelerated digitalization in many enterprises and businesses by at least 7 years. (https://www.financemagnates.com/thought-leadership/steps-in-building-a-digital-transformation-strategy-to-grow-your-business/amp/)

Another study over 1,000 large enterprises found that 67 percent are actively exploring and experimenting with new ideas regarding digital tools, compared to only 35 percent in

2018. (https://enterprisersproject.com/article/2021/3/digital-transformation-4-strategy-questions)

According to a third study, 69% of boards of directors accelerated digital business initiatives in 2020 in the wake of COVID-19 disruption, while almost half anticipate changing their organization's business model. (https://www.itpro.co.uk/business-strategy/digital-transformation/359736/the-key-pillars-of-a-digital-transformation)

Yet another study predicted that, by 2022, 70% of all organizations will have increased their use of digital technologies to drive customer engagement, employee productivity, and resilience. (https://www.itpro.co.uk/business-strategy/digital-transformation/359736/the-key-pillars-of-a-digital-transformation)

Another research paper showed that two-thirds of the CEOs of Global 2,000 companies will shift their focus from traditional, offline strategies to more modern digital strategies to improve the customer experience within a year—with 34% of companies believing they'll fully adopt digital transformation within 12 months or less. (https://www.superoffice.com/blog/digital-transformation/)

However, while many companies are progressing with their digital transformation plans, not all are necessarily implementing a new suite of digital tools and systems.

Investments in Digital Transformation

Companies are setting aside huge budgets to drive digital transformation as it has become a must for survival. Statistics show a consistent increase in such spending over the last decade.

As more and more organizations embark on digital transformation journeys, the spending on digital transformation technologies and services worldwide is expected to rise from 0,96 Trillion USD in 2017 to 2,39 trillion USD in 2024.

Research estimates the global digital transformation market will be worth $3.2 trillion by 2025.

The net global spending on digital transformation in 2018 was approximately $1 trillion. This number is expected to increase to more than $2 trillion by 2022. (https://www.statista.com/statistics/870924/worldwide-digital-transformation-market-size/)

Digital Transformation Market Revenue Worldwide from 2017 to 2022

Revenue in Trillion USD

Year	Revenue
2017	0.96
2018	1
2019	1.18
2020	1.3
2021	1.97
2022	2.3

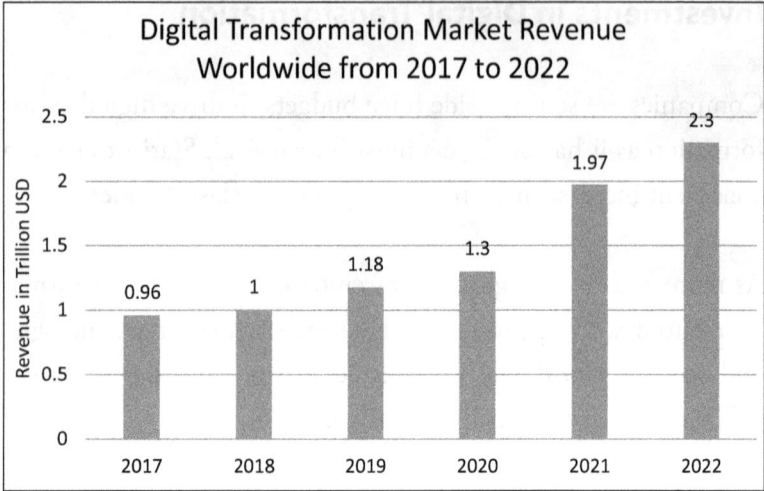

Failures and Reasons For Failure

Unfortunately, many of these initiatives fail due to various reasons. Despite the enthusiasm, digital transformation efforts continue to fail most of the time and at shocking rates.

Research indicates that roughly 70% of digital transformation initiatives fail and that more than 50% of digital transformation efforts fizzled out completely in 2018. (https://www.netsolutions. com/insights/challenges-to-a-successful-digital-transformation-and-how-to-overcome-them/)

In addition, a study showed that only 16% of employees said their company's digital transformations have improved performance and are sustainable in the long term. (https://www.netsolutions.com/insights/challenges-to-a-successful-digital-transformation-and-how-to-overcome-them/)

It is important to note that failures have been taking place in the world-class large organization across the globe. GE Digital, Ford, Xerox, Kodak, Nokia, HP, Procter, and Gamble are a few examples of a long list. Their slower pace when it came to adopting digital transformation left them open to being eclipsed by competitors as newer, more agile brands sprung up in the same industries.

Many studies and research projects have been conducted to identify the contributing factors behind such failures. These failures have cost the companies a fortune. Many of them could not survive and ran out of business.

Very often the approach and framework of digitalization turn commercial, theoretical, and complex, which further adds to the divide between the technology workforce and business leadership teams.

From our experience and exposure to research and studies, we have come across over 100 reasons why such initiatives fail, but we can summarize the most common factors for such failures:

- **Absence of a clear vision, objectives, goals, and strategy**
  This is usually the result of not understanding what digital transformations mean for the business.

- **Unrealistic budget/timelines**
  Many organizations try to do too much too soon and expect it to cost too little. It does not take long in the execution phase to realize that not enough budget has been allocated and the duration of the project prolongs. Unfortunately, in most such cases, budgets allocated to some critical activities (such as testing, training…, etc.) are cut.

- **The initiative is seen as a project and not as a journey**
  Digital transformation is not a project; it is a journey that has no destination. Once you start, you become a player in the digital world. In other words, it becomes an ongoing concern. Many failures take place because the expectation is that the project ends once you go live with your new digital platform. In fact, the journey only starts when you go live.

- **Viewing digital transformation as a technology project**
  Digital transformation is all about changing the business model using what technology offers. Without changing the business model, very little can be achieved and you end up adopting technology for technology's sake and not

linking the resources to business outcomes. This results in throwing money at buying more and more technology with no real business benefits.

- **Failure in business process reengineering**
  It is impossible to take advantage and optimize the use of technology without redefining the business processes. Through reengineering business processes, you can achieve more effective and efficient processes coupled with improved customer experience. Many companies have failed in this regard: they try to stick with the current business processes and customize the technology to outdated processes. This proves to be very costly.

- **Insufficient attention to change management (Internal culture change)**
  The term 'transformation' refers to a total change in the business model. Unless managed carefully, this change can lead to a total failure. Transformation brings about uncertainty, fear of unknowns, and risk. In addition, it requires upgrading skills through training, knowledge of new processes, new technology, a new operating model, and the introduction of a totally new culture. Therefore, huge efforts are required to manage this change throughout the digital transformation journey. In our opinion, this is the top reason for failure—and almost all failures have some element of change management attached to it.

A major aspect of change management is communication. Good communication is a must for a successful implementation. It might sound logical and simple but, in practice, it is overlooked.

- **Poor project management**
  Many projects fail due to poor project management—in other words: poor execution of the project. While all projects require strong project management skills, when it comes to digital transformation projects, the need is even greater due to the complexity, dynamic nature, and magnitude of such projects. It is quite typical in digital transformation projects to have wide variations that need to be catered for during the execution due to the fast pace of technological enhancements. Being agile and able to meet changing demands is essential.

- **Silo thinking**
  Most organizational structures are based on functional responsibilities. This leads to thinking of requirements in a siloed way. Digital systems rely on processes. If they are not fully integrated, these processes will not yield effective and efficient results. Moving away from functional thinking to process thinking has been a major hurdle in digital transformation projects. The effect of such problems can become a burden on the customer and employee experience.

- **Board and top management support**

  Leadership in any organization plays a vital role. Without the full support of the board and the involvement and commitment of the top executives, no digital transformation project can become successful. In many cases, the project is left to the CIO, and the leadership team thinks that support means providing resources and a budget. It goes far beyond that: requiring full commitment through active involvement. Decisions need to be taken in time with full buy-in. Success and failure need to be the responsibility of the top leadership team, not just the CIO.

- **Insufficient/lack of skills**

  Digital transformation requires a new set of skills, mainly in IT but also in business. Many organizations opted to upgrade the skills of internal employees without acquiring the required skills. New technologies require expertise to ensure successful execution. Training the internal team is good, but must be complemented with required expertise for a suffucient period. The same applies to the area of business as well. You must complement your business team with resources that have the necessary set of skills.

- **Data management**

  Going digital means dealing with data. Without proper data management, success becomes out of reach. This includes a definition of data, analysis, protection, and knowledge of related regulations. A key requirement is data integrity; that is, confidence that the data can be trusted.

From this core list of pitfalls come many combinations and branches that may combine to the detriment of the brand or enterprise. Who could have foreseen the seismic shift in business IT requirements over the last twenty years and even more so in the last three or four? Even our best data models would have struggled to forecast the speed and nature of change. But where some floundered, others flourished. What, then, makes the difference?

Despite many failure cases, there are many major success stories as well.

Leveraging what technology has to offer and deploying the right technology at right time and in the right way has led many organizations to succeed in their digital transformation endeavors.

Through digitalization, artificial intelligence, and capitalizing on the Internet of Things (IoT), industry leaders have set good examples and practices for successful digital transformations. The list includes McDonald's, Hasbro, Capital One, Chipotle, IKEA, Trelleborg, Microsoft, Coca-Cola, Nike's Custom Shoes, Target's Curbside Pickup Model, Netflix, Adobe, and Unilever.

The Need For Effective Digital Governance

A major question that immediately comes to mind is how can such large world-class companies allow for such failures? Why didn't they put any governance in place?

The answer is simple: all of them *have* put governance in place. The obvious conclusion is that the governance put in place has not been *effective*. An effective and practical governance framework is crucial to ensure successful implementation of digital transformation initiatives.

Governance of digital transformation is much simpler than the prevailing perception of complexity, provided a practical approach is adopted. Without an effective governance framework, the digital transformation journey may turn out to be a never ending journey.

Digital transformation is one of the streams of broader transformation and is related to digital governance. Transformation is a different way of working. It involves changes in beliefs, values, and intent. Transformation results in a shift in the organizational system and, as a result, in personal and organizational behavior.

Successful execution of any project demands effective governance. When it comes to larger and more complex projects, the need becomes even greater. Add a dynamic fast-changing technology landscape, and the complexity continues to increase exponentially.

Many governance models have been introduced to deal specifically with digital transformation. However, these models are often too complicated, and/or too technical. Such models can require significant time investments that may not be affordable by C-level executives, or even the engaged project team who is trying its best to achieve the target.

We passionately believe that for any governance model to be effective it needs to be simple and should cover all key components detailed in the book. Complex models are obstacles on their own. Therefore, we have come up with a practical governance model targeted towards the Board of Directors and CEOs.

Components of Digital Governance

Digital transformation is a journey whereby business processes are conducted in a smarter and agile manner with automation and integration embedded in the processes.

Effective digital governance can lead to successful transformation if all components are present and used effectively. Those components are:

Digital Strategy and Execution

Before embarking on defining your digital strategy, you must first define a clear vision of where your organization wants to be in the next few years.

This vision must be in line with the purpose and mission of the organization and outline the expected business benefits of the digital initiative in detail.

Once the vision is in place, a digital strategy can be defined that sets the plan and roadmap on how you are going to realize the set vision.

Supporting ecosystem

Any initiative requires a relevant and agile ecosystem for successful execution.

The ecosystem consists of the organization's stakeholders such as employees, shareholders, customers, regulatory and government authorities, partners, and vendors.

Change Management

When it comes to digital transformation, the amount of change that will be introduced throughout the journey will impact all aspects of the business (processes, business model, people, systems, etc.).

Therefore, special attention must be paid to change management and a clear strategy must be defined to deal with this challenge.

Standards, Procedures, and Policies

Fortunately, many globally accepted frameworks and standards exist that guide how to address a certain topic.

They provide best practices that are regularly updated to address any gaps. They also help in defining procedures and processes and ensuring adequate control.

Considering that there are so many standards available, it is important to select the right one that matches your need.

Performance Management and Control

One of the biggest issues in the digital world has been project performance management.

It is important to ensure that adequate measures are put in place so that the digital transformation journey is on track in terms of cost, schedule, quality, and desired outcome in terms of functionalities.

Cyber Security

In many organizations, cyber-attack has been on the top of the risk register.

The damage from cyber-attacks can be quite expensive, not only in financial terms and operational efficiencies but also in reputational damages.

Embarking on a digital transformation journey means introducing new technologies to the organization and customers and other external parties. This puts a lot of burden on ensuring the comprehensive protection of systems and sensitive data. Therefore, significant focus has been put on cyber security.

Emerging Technologies

New technologies are introduced daily and, as an organization, you need to be able to evaluate them effectively.

While new technologies bring about new features and benefits, you cannot keep introducing them too often as they are expensive and introduce some issues and risks (integration, security, complexity, skills, and change management).

On the other hand, you should not close your eyes and simply hope to maintain the status quo. Lagging too far behind or adopting change too slowly may be detrimental to business. This could result in losing ground and business to competitors or failing to ensure continuity of technical support and security issues.

Therefore, it is important to strike the right balance and maintain an optimum position when it comes to your technology landscape.

Talent Management

With all new developments in the technological space, the need for new talents and skills has posed lots of challenges.

Organizations need to come up with a strategy to attract, retain, and develop new skill sets that are required in the digital world.

Data Governance

The increase in the use of digital solutions has led to an unprecedented rate of increase in data. Such huge amounts of data must be dealt with effectively to create value for the organization.

To ensure that data is secure, available, and used properly requires implementing a data governance model that deals with those challenges.

What's next?

With the growing number of digital transformation initiatives, high failure rates, and large investments as indicated earlier, it is a must to ensure that effective governance is in place so that failure is not an option. This is the purpose of digital governance.

In the next chapters, we will tackle each of the above components of digital governance in detail. We will provide relevant statistics, real case studies, and our own experiences.

1

STRATEGY AND EXECUTION

Recipe to success – strategy first and then execution

Would you dare board a flight if you were aware that the crew does not have the institutionalized checklist to be verified before take-off, cruise, and landing? The answer is an obvious No. Similarly, in any digital transformation journey, a clear strategy is critically important before embarking. A world-class strategy definition is important but equally important is aligned execution of this strategy.

Unfortunately, many strategies put in place are unclear, incomplete in terms of their coverage, or ineffective.

One of the key challenges identified in many studies is the lack of a defined strategy that is aligned with business strategy. (https://www.panorama-consulting.com/digital-transformation-challenges/)

A CIO Survey conducted recently points out that only 41% of companies have an enterprise-wide digital strategy, and only 18% of companies rate their use of digital technology as "very effective." (https://www.impactmybiz.com/blog/blog-5-change-management-strategies-for-digital-transformation/)

Vision and strategy combine to create one of the key pillars of a successful digital transformation journey. (https://headspring.com/2020/12/10/the-5-pillars-of-digital-transformation/)

Digital transformation is a major event in the life cycle of any organization and hence a significant milestone in its business journey. Such a major milestone must be preceded with meaningful and insightful strategy, coupled with a solid and detailed execution plan that promises success.

Defining best-in-class strategy without successful execution does not do any good. Another major source of overall digital transformation failure is a failure in execution.

Therefore, we have combined strategy and execution under one chapter.

This chapter will go into detail exploring how to define a digital strategy and execution plan that can lead to success, the various aspects that need to be considered while defining the strategy, and the best practices to execute and successfully implement such a strategy.

First Things First (Purpose and Vision)

Before defining a strategy, the organization must define its mission, purpose, and objectives for embarking on a digital transformation journey. Without a business purpose, which is clearly articulated, how would the organization define which direction to go? Unfortunately, in many cases, the absence of a purpose has been the main reason for failure.

You might be wondering how organizations embark on such a costly, complicated, long journey without a purpose. The answer is that often they define a purpose that is either too generic or is unclear and cannot be understood.

Here are a few examples of purpose statements that do not serve any clear business objectives:

- We want to be present in the digital world.
- We are a large enough company and must be present on the internet.
- The objective is to be at the forefront of technology.

On the other side, here are some examples of good purpose statements:

- We would like to transform our business and be able to serve our customers digitally.
- The purpose of our digital transformation journey is to reduce our processing time, improve our customer services, and enable our growth across the globe.

To complete this, it is important to have a vision for where you would like the organization to be in the future. For example, a good vision would look something like this: "We would like to increase our digital sales to become 50 percent of our total sales by 2025," or "we would like to grow our business by 30 percent over the next 5 years through the introduction of digital channels."

In short, the vision and mission statements are prerequisites of the digital strategy.

What is Digital Strategy?

Simply put, a digital strategy outlines how the business is going to run digitally, and how it is expected to achieve the organization's vision, overall business objectives, and goals.

Since digital transformation is a journey, the strategy outlines how it complements the existing business strategy. It provides a detailed plan that includes all steps, components, technology, tools, skills, budget, cultural changes, and other considerations that are needed to achieve the overall corporate vision.

In other words, it provides a clear direction and a long-term plan on how the digital vision of the organization is going to be achieved while ensuring uninterrupted business operations during the journey by aligning it to an overall business strategy.

To reiterate: the digital strategy should always be aligned with business strategy.

Digital strategy must not be confused with technology. Digital strategy is not only about technology (although technology is one of the core components). It must include various other components, like customer need, human resources, organizational culture, budgets, and change management information.

Digital strategy must be defined by the highest level in the organization (i.e., Board of Directors). We have seen many cases where digital strategy is defined by the IT department thinking it is all about technology.

Digital strategy is not a static single polar document, which can be created in isolation without reference to business objectives. The strategy needs to be agile and not static. The transformation program may commence based on the developed strategy, but the strategy itself may undergo revisions for best alignment, thereby making it agile.

The reasons for revisions could be technology updates, business needs changes, market conditions, or other arising factors.

One of the organizations we know closely had drawn out a digital strategy roadmap for the next 3 to 5 years. They kept it agile, which gave them flexibility to keep updating and revisiting to keep it aligned with business strategy and ever-changing technology and corporate landscapes. This resulted in a release of updated versions of the strategy at reasonably regular intervals. The main motto was "Frictionless IT". This has resulted in great success.

We also have seen some cases of digital strategies that are single paragraphs and are far too abstract.

A successful digital strategy must be detailed enough to include every aspect of the business that will be affected by the digital transformation program. It must at least contain the following in detail:

- Current state/business case
- What will be achieved
- Methodology, technology, approach

- Skills requirements
- Timeline
- Investment
- Expected benefits, both tangible and intangible

Why Do We Need a Digital Strategy?

Having talked about what a digital strategy is and what it isn't, let us explore why we need a digital strategy.

Failing to plan is planning to fail—and without clear direction, success cannot be achieved. A digital strategy enables the organization to outline and create a clear roadmap to achieve strategic goals. It also helps in setting benchmarks and comparing future progress against those.

The strategy should be able to provide you with an advanced glimpse of what success will look like and what challenges are expected so that your organization can prepare to deal with them as they arise.

The absence of a digital strategy can lead to:

- Loss of direction (where the organization is heading)
- Opportunity loss (that the digital world can provide)
- Losing out on technological advancements

- Trailing behind competition
- Loss of customers – customers today expect businesses to have digital channels. If such channels do not exist; they need to know when and how they will be available.
- Suboptimal processes, lack of integration
- Missing buy-in from your organization

Therefore, it is imperative that a solid digital strategy is in place and understood by the entire organization.

Pillars of Digital Strategy

There are three pillars into which all components of the digital strategy fall: people, processes, and technology. Let's look at each one in detail:

People

This is the most important pillar of the digital strategy. Most failures are related to the human factor in one way or another. Leadership plays a vital role in the strategy from formulation, execution, revision, and completion.

All of the top leadership must take ownership of the strategy, and not the head of the IT department. This means that full accountability must be taken at the highest level and by all executives.

The amount of change that a digital transformation journey brings to the organization cannot be understated. Therefore, a detailed program that ensures alignment of all employees to the new reality must be in place and executed successfully.

A timely upgrade of skills is also necessary to ensure the adoption of new digital processes. In short, a new digital culture must be created in the organization.

Customers also need to be taken care of. With a new digital experience, customers need to be looked after carefully. Customer experience is arguably the most common reason for the failure of digital transformation initiatives.

Processes

No digital transformation journey succeeds without revisiting the processes and redefining them to ensure that they are optimized and fully integrated.

This is usually done through a business process reengineering exercise (BPR). Without optimizing the processes, the journey can prolong, and the customer experience can become poor.

It is important to keep in mind that optimizing the processes should not be done at the expense of control. Striking the right balance between agility, discipline, compliance, and control is the objective.

Processes can be integrated along the supply chain to gain further optimizations. This can be achieved by integrating with suppliers, business partners, customers, stakeholders, authorities, and any relevant parties.

Processes can be measured in terms of efficiency through well-defined key performance indicators (KPIs).

Most organizations factor in the feedback of the customers to ensure continuous improvements on the processes.

Technology

Digital means the use of information technology. Therefore, it is vital to select technologies that are fit for purpose and provide the right solutions that are scalable, mature, secure, and reliable.

Today, there are many emerging technologies that make the choices tricky. Technologies such as cloud, artificial intelligence, data analytics, robotic process automation, augmented reality, virtual reality, mixed reality, and Industry 4.0 are just a few examples.

We will shed some light on some of the key emerging technologies later in the book.

Architecting the right technology landscape is of utmost importance. Select what is needed and do not go after becoming a technology house.

Whichever technology you use, you must ensure seamless integration of your systems so that users have a good experience using your systems.

Data management is another technology component. You must ensure the integrity of data so that customers trust your systems. In addition, the security and privacy of all data is a must. Many regulations deal with data breaches.

Characteristics of a Successful Strategy

All the above pillars sit on a governance layer that ensures successful implementation of the strategy by ensuring adequate project management, defining adequate key performance indicators, updating the strategy regularly.

For a strategy to be successful, it must be:

1. **Clear:** The strategy must be built on clear vision, goals, and objectives.
2. **Relevant:** It must be relevant to the organization's overall business objectives.
3. **Comprehensive:** Strategy must address all stakeholders' requirements, and must be complete in terms of addressing all business areas and processes.

4. **Principle-Based:** It must be based on clear principles such as data integrity, security, integration, fairness, and transparency.

5. **Agile:** It must be agile and flexible enough to cope with customer needs, changing business environment, and technological advancement.

6. **Scalable:** Should provide scalability upwards and downwards as and when needed.

7. **Realistic:** It must be realistic and executable through available resources. Surely you can stretch the target and expectation, but it should not go beyond what is possible.

8. **Risk-aware:** It should identify potential risks and put plans in place to mitigate and manage them.

9. **Best Practices:** It is best to take advantage and deploy a best practices approach.

Digital Strategy Execution

Once the strategy is defined, the execution phase starts. To execute the strategy, there are key elements that allow for a successful implementation journey.

From our experience, we have come up with the following approach that provides a checklist to achieve the desired results:

1. **Define a clear roadmap**

   Digital transformation is a long journey. This journey will consist of many projects and stages. Therefore, it is important to define a clear roadmap that outlines every stage.

   For every stage, identify the expected outcome, the start and completion date, the resource requirements, and the rollout plan.

   In addition, clearly define how to measure success.

   This roadmap needs to be updated regularly to cater to new requirements and any necessary changes.

2. **Put customer experience at the forefront of everything**

   The key purpose of the digital transformation journey is to provide customers with a better experience. Designing and implementing systems that are internally focused without putting customers at the forefront could prove to be quite costly.

3. **Identify all resources required**

   As stated earlier, the journey is a long one and will require many diverse resources.

Apart from the financial resources, you must have the right skills for the project. Different skills are needed at different stages of the journey. Such skills are sometimes costly and rare and must be provided at short notice.

In addition, technological and infrastructure requirements need to be designed and architected adequately to ensure the smooth running of the system with high availability and solid disaster recovery and business continuity plans.

In cases where the integration of third parties is involved, it is important that they provide you with the right information.

4. Define and select standards and frameworks
During execution, it is important to select which standards and frameworks you are going to use and use them consistently throughout the project lifecycle.

This will provide you with many benefits when it comes to project control, benchmark, and guidance, as such standards have proven track records.

5. Ensure alignment with applicable laws and regulations
Entering the digital world requires that you abide by regulations where and when they apply.

Different countries, industries, products, and services have different applicable regulations. A checklist should be created for all relevant applicable regulations for each category.

6. **Use effective and agile methodology that suits best**
Many agile methodologies exist that can be used in project implementation.

Since digital systems require frequent updates. We recommend that you use the agile method that suits you best. This will enable you to respond to market needs quickly.

7. **Ensure strong PMO is in place to ensure success**
It is a well-known fact that IT projects are difficult and without a strong project management office, you might not be able to control the project adequately from scope management, cost, quality, and time control.

8. **Define key milestones and KPIs and measure progress and results**
Monitor the project progress continuously. This can be achieved by defining some key performance indicators that can be reported through a dashboard that is updated all the time.

Updates should be provided at least weekly. The KPIs should include all aspects of the project that need to be controlled.

9. **Ensure that desired level of quality is achieved**

    It is quite common amongst IT projects that when a delay occurs in the project execution, the quality testing duration is cut short.

    This would result in an outcome with many teething problems. Never shorten the testing time, as this will lead to quality issues.

10. **Do not forget to address change management**

    Although the topic of change management is talked about so much, the plain truth is that in almost all cases, it receives insufficient attention.

    This is also true when it comes to training the system users.

    Quality and timely training must be provided to ensure that the user is familiar with the system.

In one very large software implementation program, which included process review and transformation, a dedicated change management team was formed that worked very closely with the main project team. The project adopted a 'Train the Trainer' program approach, through which functional users were trained by professional trainers and software consultants.

The organization was surprised when they noticed that some of the users, who were very shy in their regular life and would hardly ever open their mouth, turned out to be very effective trainers and guide the end-users. The organization started getting much stronger. This was the effect of classic change management.

Beware of Potential Failure Hotspots

There are many potential failures that can take place. The following list highlights some of the known challenges that we have come across in our experience:

Lack of dedicated IT skills. According to one recent survey, 54% of organizations reported that skill shortages (mainly in the areas of Cyber Security, Technical Architecture, Enterprise Architecture, and Advanced Data Analytics) were holding them back from pursuing their transformation goals. (https://www. panorama-consulting.com/digital-transformation-challenges/)

Lack of organizational change management. Outdated organizational structures, inefficient workflows, and rigid leadership styles can all impede digital transformation success.

Evolving customer needs. Customer needs change continuously because of changing market conditions. Failure to cater to these needs in time has resulted in many failures.

Lack of a defined strategy. We have highlighted earlier the need for a clear well-defined strategy. It comes as no surprise that the absence of strategy is a formula for failure.

Budget concerns and constraints. This comes because of short-term focus. Sufficient budgets must be allocated. Most failure cases have either underestimated the budgets or have budgeted for the first phase and have not considered long-term requirements.

Ineffective data management. To serve customers effectively, a data strategy should be in place that enables the organization to know the customer well and hence fulfill their needs. Many failures are due to lack of data and information.

Inefficient Business Processes. This is the result of not changing the business processes as mentioned earlier. Without a proper business process reengineering, it becomes difficult to become efficient and effective. After having defined a successful digital strategy, we need to talk about the support ecosystem for robust execution.

An ambitious program like a digital transformation cannot and must not commence without a purposeful and effective strategy.

This endeavor requires serious change management initiatives, without which many digital transformations have failed to yield desired results. In the next chapter, we will review all of these aspects in detail.

2

THE SUPPORTING ECOSYSTEM

Your companions are the ones who will make your journey a Success

Can you imagine the earth without air and water? The obvious response is No. Similarly, digital transformation is not conceivable without a supporting ecosystem, and, without change, no transformation takes place.

Digital transformation journey cannot be successful without a strong supporting ecosystem. In this chapter, we will explore what the digital ecosystem consists of. We start with defining the term and then identify the major and key players in this system, the roles they play, the responsibilities they have, their expectations, and how to manage them. According to McKinsey, digital ecosystems today power 7 of the world's 12 largest companies by market capitalization. (https://www.mckinsey.com/business-functions/strategy-and-corporate-finance/our-insights/the-strategy-and-corporate-finance-blog/if-youre-not-building-an-ecosystem-chances-are-your-competitors-are)

To complete the picture, we will dedicate some space for change management. The term transform, by definition, means to move from one state to a completely new state. Failure to manage change may lead to failure of Digital transformation. Therefore, it is important to pay special attention to change management.

What is the support ecosystem of digital transformation?

The ecosystem of digital transformation is a purpose-based, seamlessly connected, and interrelated system of entities working for a common purpose. The composition can be an intelligent mix of both internal and external entities, which could be tangible or intangible spanning across human resources to digital knowledge and domain solutions. This network is made up of many players such as employees, vendors, customers, partners, technology applications and solutions, third-party service providers, competitors, and all related & relevant technologies.

The internal ecosystem consists of employees, shareholders, subsidiaries, organization culture, systems, technologies used, knowledge systems, policies, and procedures. The external ecosystem consists of customers, partners, suppliers, competitors, shareholders, government, and regulatory authorities.

Ecosystems can also be broadly classified as in-house ecosystems and partnership-based ecosystems. Examples of in-house ecosystems are internal knowledge, products, and tightly integrated services—whereas partnership-based ecosystems components consist of the organization partners with other internal and external players in the digital domain such as suppliers, business partners, and even competitors. This way, the organization can provide more products and services. With the digital domain expanding rapidly, we see a growing trend moving towards the partnership-based model.

Benefits of the Ecosystem

A research study conducted by McKinsey shows that an emerging set of digital ecosystems could account for more than $60 trillion in revenue by 2025, or more than 30% of global corporate revenue. https://www.mckinsey.com/ business-functions/mckinsey-digital/our-insights/how-do-companies-create-value-from-digital-ecosystems)

Ecosystems bring along with them many benefits to the organization. Here are some of the major benefits:

- Ecosystems help to embrace and retain digital domain knowledge (both internal and external). This knowledge will enable the organization to become more agile in responding to macro-level economic changes and global issues, changing market needs, and capitalizing on market and growth opportunities. This can be achieved by making faster and better decisions building on the knowledge acquired. Internal knowledge helps to improve processes, policies, procedures, and skills, while external knowledge provides relevant and timely market-related information. The recent COVID-19 pandemic has proven the importance of being agile in the digital world – for example, many organizations could change to a different workplace model, shifting from working from the office to working from home in a rapid manner. The organizations that could not carry out such a shift quickly suffered.

- Ecosystems allow the organization to benchmark against best practices, competitors, partners, and other influencing factors. This benchmark data provides the organization with information regarding where it stands in comparison to other similar organizations.

- They also enable the organization to better define KPIs. By defining the right KPIs, the organization can evaluate its performance much better and focus on what really matters for the business to perform better and to achieve its strategic objectives and targets, both short and long-term.

- Ecosystems provide access to case studies (successful and unsuccessful). In a fast-moving digital world, it is vital to understand why certain initiatives succeed and why others fail. Building a strong ecosystem helps in obtaining such information from partnerships so that the organization builds on what leads to success and quickly identifies what causes failure.

- Your business can also stay up to date with the latest developments in technology. Suppliers play a very important role in the digital ecosystem, as technologies are getting enhanced at a rapid pace, and the right relationship with suppliers can ensure that your digital landscape does not end up becoming obsolete. It is necessary to keep an eye on technology enhancements. This can be achieved by ensuring that key suppliers and technology players are an active component of the ecosystem.

- The right ecosystem can also enhance customer loyalty and satisfaction. This can be done by enhancing current product and services offerings through the partnership programs rather than building new ones from scratch.

The COVID-19 pandemic is a core example of how fast an ecosystem can change. The world was largely unprepared for the speed with which the global economy shifted. Suddenly, stakeholder needs lurched in a new direction and the organizations that survived this mass migration to a new way of thinking about work and business were the ones who survived. We believe that these were also the ones who had a firm idea of what their ecosystem consisted of and how fast it was changing. It was not taken for granted.

Major Players in the Ecosystem

There are many players in the digital ecosystem and to make the ecosystem effective, each player must be given adequate space and attention to ensure that it contributes timely and positively. Here are some of the key ones:

- **Customers:** Without customers, no business survives. To ensure that your customer needs are fulfilled all the time in an effective, efficient, friendly, and timely manner, you must keep them actively engaged in a continuous dialogue. Customer needs and demands change quite rapidly and failing to meet such needs and demands will simply result in your competitors taking away market share from you. Meeting this demand requires knowledge of what customers expect and using the ecosystem to proactively address their needs.

  One of the most important aspects is the customer experience of using the digital systems. Most successful digital transformation initiatives have put customer experience as the top key success factor and engage their customers in developing systems even during the testing phases. Make sure that customers are an integral part of your digital transformation journey.

- **Suppliers/business and technology partners:** Suppliers can provide you with the means of harnessing business agility to scale up and down quickly. They also can help you offer new products and services quickly. Business and trade partners also play a key role, whereby through the ecosystem, combined services and products can be developed quickly and hence, expand the portfolio, capture emerging opportunities, and increase market share. Many organizations today have integrated their systems with those of their key suppliers whereby the visibility of the whole supply chain has become real-time. Such information can be used to provide customers with needed and useful information. A good example would be Amazon and shipping companies.

- **Technology providers:** Technology is what makes everything happen; it also keeps improving, introducing new products, software, and services to enhance product and services offerings, improve customer experience, reduce costs, security, and improve processes. Please note that whatever technologies an organization use can become obsolete, insecure, and costly to maintain—and must be replaced quickly. Therefore, it is important to keep up-to-date with technology trends through strong relationships with key technology providers. This way, you can ensure an acceptable level of investment protection, a solid and secure technology landscape, and access to required technology skills.

- **Employees:** The saying that employees are the most important assets that organizations' have holds true in the digital world. It is the knowledge that those employees gain across the entire digital and process domain that enables them to make the internal processes robust, efficient, agile, scalable, and dynamic. The digital footprint provides employees with information that enables them to fulfill the dynamic market, process, quality control, and regulatory requirements in an optimum way.

Ecosystems From a Different Perspective

As IT Pro states in their article published in association with O2 Business (https://www.mckinsey.com/business-functions/mckinsey-digital/our-insights/how-do-companies-create-value-from-digital-ecosystems), the digital transformation is based on 6 main factors:

1. People
2. Data
3. Process
4. Optimization
5. Innovation
6. Feedback

Each of the above factors influences the outcome of the digital transformation program. An evolving workforce with varied working styles along with mixed working patterns—like work from home, work from an office, work from both, or work from anywhere—calls for a trusted and rugged technology and infrastructure.

Innovation can be achieved through the right combination of people, process, and technology. Another golden rule is to optimize the processes before digitalizing them. Digitalization without optimization will mean automation of legacy processes, which might not yield the desired outcome.

How to Build an Effective Ecosystem

Ecosystems are not static; rather, they are dynamic, continuously changing, and built over time. For an ecosystem to be successful, it needs to be managed well through effective knowledge management systems, project management, and the use of collaboration tools and other solutions. When the ecosystem components are interoperable, they bring great value to the organization. A robust framework and implementation plan for recognition and rewards is a must. The exiting strategy and post-implementation rehabilitation is an important consideration. Here are some important factors that make the ecosystem effective, contributing positively to digital transformation:

1. Create a culture of collaboration within the organization and with other ecosystem players (internal and external). Building such a culture requires strong leadership, transparency, and a win-win thinking approach.
2. Get rid of internal silo systems and processes. Boundaries between business units or internal departments must be removed.
3. Ensure that employees are fully engaged, as the journey requires that everyone is on one page. Everyone must be involved in creating process efficiency, actively building a knowledge base for the organization.
4. Acquire and implement technological tools that help you build the knowledge, encourage collaboration, and result in integrated systems with enhanced capabilities that support the ecosystem.
5. Ensure effective communication and documentation about the ecosystem. After the successful formation and deployment of all or work in progress components of the ecosystem, it should be documented and communicated

properly. Each component of the ecosystem must be fully aware of their roles, expectations from them, KPI, KRA, etc. Equally important is to make the organization aware of it. Performance management of every component will be an important consideration.

The ecosystem formation process and organizational awareness of this require concerted effort. For non-historical tangible in-house and external components of the ecosystem, the following approach may be adopted:

1. During the pre-project period or just at the start of the digital transformation project, develop and agree on a **project charter**. This charter should be accepted and signed by the top management, preferably the CEO of the organization. An important component of such a project charter will be a **project organization structure**. The project organization shall consist of the details of tangible internal and external components.

2. After defining the project organization, which mentions the tangible components, the project charter details the **position charter** for each such component. The position charter mainly consists of the following:

   a. Role / Purpose of the ecosystem component (for example Project Sponsor Business Process owner or Super User)

   b. Job responsibilities

    c. Required skillset, experience, and required behavioral characteristics

    d. Criteria for performance measurement

    e. Project-related reporting arrangement

    f. Expected time commitment

3. The above parameters should be taken into consideration while selecting these entities. We recommend avoiding nominations for entities like super users, who are expected to be dedicatedly associated with transformation projects. Such entities should be identified and selected based on the required skills set and performance criteria.

4. After the formation of the project organization, its formal announcement with adequate importance and due recognition should be made in the organization.

The clear realization of the need of each ecosystem component, sourcing and deployment of all such components of the ecosystem, and their performance management is key to the success of any digital transformation program. Equally important is organizational communication and ensuring that maximum value realization is achieved from all components of the ecosystem.

3

CHANGE MANAGEMENT

You don't have to see the full staircase; just take the first step

"Never underestimate the magnitude of the forces that reinforce complacency and that help maintain the status quo."

— JOHN P KOTLER

It has become rather repetitive talking about change management as it has been talked about for years, especially when it comes to IT projects—and many books that focus on change management have been published. Unfortunately, most organizations do not seem to attach enough importance to it.

Change management is the cornerstone of any successful IT project. When it comes to digital transformation with so many internal and external parties involved, it becomes even more important and more challenging.

Change management refers to a set of systems and processes that enable the organization and all the ecosystem layers to make the transition from the current state to the newly desired state effectively and efficiently. This transition must be irreversible (i.e., one-way) so that everyone in the organization uses the new systems and processes. Going back to an old system can pose a threat to process effectiveness.

The change must start from the top leadership team, as they should play as the role model and others will follow. First, ensure that change is necessary and desirable. Any type of disruption should be minimized. Communication should be promoted, and it must be recognized that change is the norm and not an exception. The goals of change management should be enhanced ROI, competitive environment creation, and the chance to empower and energize the employees. As Sanjay Brahmwar, CEO of Software AG mentions in his post in Software AG blog: a clear cultural change strategy is imperative in successful digital transformation projects. The "why" of the change and efforts towards clearly communicating the why is more important than "what" and "how" of the change.

Poor change management is often one of the main factors for digital transformation project's failure. According to McKinsey & Company, 70% of change management efforts fail due to employee resistance and lack of management support. (https://planergy.com/blog/digital-transformation-change-management/)

There are many challenges facing change management:

- **Fast-changing technology landscape**
  In today's world, technology is advancing at a very rapid pace. To be able to adopt such technologies, organizations need to be able to change quickly. With newly introduced agile methodologies, the need to fast adapt to market/customers/regulations/etc. has become a requirement.

- **Organization culture**
  It is difficult to try and change an individual's behavior, and when it comes to an entire organization, the task becomes quite complex. Many norms will have to change in quite a short period.

- **Unsubstantiated external influences**
  Quite often, there are external influences that make it difficult to change. This could be in the form of parent

organization, laws and regulations, or industry norms, among others.

- **Consensus does not always work in large digital transformation programs**
  Getting the top leadership aligned is not easy. Many organizations revert to voting and other methods to arrive at a consensus when it comes to decision-making. In many cases, this leads to the loss of top management support.

How to ensure successful change management

To ensure that successful change management takes place, we recommend the following guidelines:

- Define a clear change management strategy and use one of the well-proven change management methods (such as ADKAR).
- Ensure that the full leadership team is on board and actively participates and supports the digital transformation journey. They must walk the talk and set a good example.
- Communicate the digital transformation journey's objectives across all levels of the organization. This communication needs to take place throughout the journey.
- Create a sense of urgency. It is do or die for everyone.

- Minimize disruption during project implementation. Going through multiple disruptions brings about multiple stages of change and makes it more difficult to adapt.
- Have regular roadshows to keep everyone updated and engaged.
- Celebrate success at different stages of the project.
- Recognize that this journey is difficult for everyone and deal with change management issues as they arise. Do not ignore them.
- Investigate the silence thoroughly: silence might not be always golden.
- Persuade the employees that change brings in opportunities and it's not always done to take away their jobs. At the same time, you or anyone else leading a change initiative must avoid feeling sentimental about inevitable job losses.

It will be worthwhile attempting to get further clarity on the difference between change and transformation. This is very well explained by Erika Flora in her article in Beyond 20: "A snake shedding its skin is a change. Old skin is outgrown and is of no use and the snake works to get rid of it. A lot of what we do as organizations to get better or improve are constant improvements or changes. A transformation, however, is when a caterpillar becomes a butterfly." (https://www.beyond20.com/blog/digital-transformation-with-organizational-change-management-and-itil-4/)

Based on our experience in different organizations and projects, we realized that in any organization there are three types of users:

1. Leaders
2. Followers
3. Betrayers

Leaders are the group of users who are always willing and ready to take the lead in the projects and implied changes. This is a group of positive employees, ready to move forward with full confidence and conviction. No additional efforts are needed on such users; they are assets to any project.

Followers are a group of users, who are always there with all initiatives as supporters in a non-destructive manner. They will not take the lead on their own, nor they will be detrimental to projects. They just need guidance, and they walk with the project and accept changes.

The third category of users, what we refer to as **betrayers,** requires very careful monitoring and mentoring. This category of users would remain on a constant lookout to derail or sabotage the project and resist any change initiatives. An appropriate strategy and action plan needs to be in place to handle such users, or they might turn out to be fatal for projects and any change initiative.

During one of such projects in India, when we were in a project to computerize all the offices and outlets of a leading steel manufacturer in India, we came across a very aggressive trade union member, who was extremely against any form of computerization. He had very strong clout on the employees. The best approach was to bring him inside the "ring" – we made him the local superuser of the project and he turned out to be the best implementation support.

In another large implementation in UAE, we approached the entire change management process by assessing the organization in the following manner (this was done with the help of an external consulting organization):

The organization was analyzed through a "double perspective" approach. The first perspective was "inside looking outside," through which the organization was evaluated in light of the scope and objective of the project, whilst "outside looking inside" assessed the project through the eyes of the organization.

The analysis in the segment of "inside looking outside" primarily focused on: business practices and organization structure, organization culture and its ability to change, organizational communication, and organizational competencies. The analysis under "outside looking inside" considered the aspects of project content, project influence, stakeholders, project communication, project team dynamics/collaboration, and project team profiles.

In addition to the above, the organization was also assessed on change readiness (how ready is the organization in respect of this specific change?), change magnitude (how deep and wide we perceive this change to be?), change retrospective (how well we have dealt with and managed the change initiative in past) and team collaboration (how well we work together to make a change happen).

The above study formed the foundation of a robust change management strategy in that organization and for a big-size project.

Change management is a complex and necessary component of any digital transformation program. There is a universally associated phenomenon with any change management initiative: resistance to change. It is basic human nature to feel more comfortable with the status quo. Any resistance to change, howsoever far-fetched, must be dealt with seriously. The build-up resistance may turn out to be fatal to the project, if not addressed properly.

Change management is such a vast topic that it may require a full book to do justice to the subject. But, since the topic of this book is digital transformation and change management being a critical component for any such journey, we attempted to highlight the importance of change management through this chapter.

4

FRAMEWORKS, STANDARDS, AND THE 3 PS

(Processes, Procedures, and Policies)

Consistency and quality can only be achieved with discipline and style

Imagine a city without any established traffic regulations, and hence vehicles are free to decide which side of the road to drive on at their will. Clearly, this would cause chaos!

Meeting applicable standards is the minimum you need to stay in the game: exceeding them will empower you to compete with the best.

Many frameworks and standards exist that have gained wide acceptance across the globe following encouraging and successful results.

These standards and frameworks can help you in your journey whether the organization is in the project execution phase, post-implementation phase, or in a stable state.

By selecting the most appropriate frameworks and standards that apply to a particular business, you will increase your chances of success in implementation and having a solid stable business operation.

To ensure quality and consistency in any service delivery or product development, it is mandatory to practice a set of relevant pre-defined steps and frameworks in a structured and systematic way. Such predefined steps and frameworks are complemented through policies, procedures, and processes that ensure strong governance in using the standard or framework.

Through a process of evolution and maturity, the IT industry has come up with several established and tested standards and frameworks for relevant digital transformation objectives.

In this chapter, we will talk about widely used standards and framework, and their benefits, need, and usage. We will also shed some light on what policies, processes, and procedures are in the context of a digital operations model.

What Do We Mean By Standards and Frameworks?

A standard is a model or guideline towards a topic, product, process, or procedure. A guideline usually becomes a standard after it has gained global acceptance and has proven its usefulness.

No standard is perfect. Therefore, all are reviewed and enhanced continuously as time goes on. Since technology changes rapidly, quite often new standards are also introduced to fill in the gaps in older versions or to address new requirements due to changing business scenarios and other dynamic factors.

While many frameworks and standards exist in the Information Technology environment, only a few of them have gained global acceptance and are common. Such standards provide structured approaches and proven methods towards managing strategy and execution.

Policies, processes, and procedures need to be put in place to ensure an effective operating model that leads to desired goals.

A **framework** can be described as a structured approach towards addressing your needs and achieving the desired outcomes. Frameworks are used to provide guidelines and best practices.

The difference between a **standard** and a **framework** is that a framework provides you with guidelines on how to go about doing a certain task or project. It is supposed to represent best practices and you are not obliged to use it.

A **standard**, on the other hand, is a well-defined set of clauses and requirements. Some standards are required by law and regulation and you must follow them, while others might be optional.

Certifications and audits take place against standards, but not frameworks.

What Are Frameworks and Standards Used For?

Frameworks and standards are used for many purposes:

- Marketing and customer confidence purposes. Those address quality, security, or in some cases, a specific profession, or fields such as health or education.
- Enhance internal control, effectiveness, and efficiency. Those address the internal processes in detail.
- Ensure proper project management

- Information and Cyber Security
- Enforced by authorities to ensure that the systems and solutions are in line with the laws and regulations and to protect the stakeholders' information and privacy

In summary, frameworks and standards can be used for various purposes and are updated regularly to include new developments in the areas they cover.

Benefits of Using Frameworks and Standards

There are many benefits to using frameworks and standards:

- Provide confidence to your customers and stakeholders and reduce risk in the sense that they have been widely tested, accepted, used, and proven to work.
- Ensure consistency in the processes, which leads to process stability in systems with effectiveness and efficiency.
- Offer benchmarking opportunities to compare your organization with others who are using the same or similar standards throughout your digital transformation journey.
- Ensure adherence to laws and regulations. Usually, standards address global and local laws and regulatory requirements, and therefore, help you abide by them. In most cases, laws and regulations establish their

requirements based on some of the well-established standards.

- Provide a reference for auditing and control. Standards are easily auditable and there are certifications for auditors that allow them to audit any organization against those standards.
- Facilitate KPI definitions and measurement and allow you to better define and measure KPIs and helps manage strategy, risks, priorities, execution, and critical issues

In summary, frameworks and standards provide you with an approach that has been proven, can be benchmarked, and audited.

What Are Common Frameworks and Standards?

The most common standards and frameworks that may be considered:

1. **COBIT (Control Objectives for IT and related technologies):**
   COBIT is an IT management framework developed by the ISACA (Information Systems Audit and Control Association – a not-for-profit organization) to help businesses develop, organize, and implement strategies around information management and governance. It was first

introduced in 1996 and has been updated many times. The latest revision was made in 2019.

It basically addresses IT governance, control, and management areas.

Many organizations use COBIT as a reference to see how mature their Information Technology department is. For example, the Dubai government uses COBIT to rank the maturity of IT departments in all its entities.

2. **ISO/IEC 27001 – Information Security Management System (ISMS)**

An information security management system is a systematic approach to risk management, containing measures that address the three pillars of information security: people, processes, and technology.

The purpose of the system is to protect the organization's assets.

The ISO/IEC 270001 family of standards, also known as the ISO 27000 series, is a series that provides best practices to help organizations improve their information security.

It was first introduced in 2005. The latest revision was made in 2013.

3. **ITIL – ISO/IEC 2000 (Information Technology Infrastructure Library)**

   ITIL is a framework of best practices that helps to deliver high-quality IT services.

   The approach ITIL takes is intended to combine processes, people, and technology to support service delivery, evolution, and maintenance for end-users or customers.

   ITIL covers services in all areas of IT including such as applications, software, infrastructure, security, and user support.

   ISO/IEC 20000 is the international standard specifically for IT Service Management that is based on ITIL principles. It describes an integrated set of management processes that form a service management system for the effective delivery of services to the business and its customers.

   ISO 2000 was introduced in 2005 and the latest updated version was released in 2018.

4. **PCI DSS (Payment Card Industry Data Security Standard) – Used for card payment.**

   As the use of the internet has been growing at a rapid rate and more and more payments are processed through the internet, there has come the need to ensure the security of such transactions.

The Payment Card Industry Data Security Standard (PCI DSS) is a set of security standards designed to ensure that all companies that accept, process, store or transmit credit card information maintain a secure environment.

An independent body created by Visa, MasterCard, American Express, Discover, and JCB, the PCI Security Standards Council (PCI SSC) administers and manages the PCI DSS.

Surprisingly, the payment brands and acquirers are responsible for enforcing compliance, rather than the PCI SSC.

Version 1 of PCI-DSS was released in 2006 and version 4 was released in 2021.

5. **GDPR (General Data Protection Regulation) – EU's Data privacy and protection law.**
   Lately, there has been a lot of controversy on data privacy. With the popularity of social media platforms such as Facebook, Instagram, and others, protection of personal data has been demanded—and many countries have developed data privacy laws to protect individual and business information.

   The European Union (EU) has passed a law called "General Data Protection Regulation". This law sets guidelines for the collection and processing of personal information from individuals who live in the European Union (EU) countries.

The law was introduced in 2016 and became mandatory in 2018 at the time the latest update of the law was introduced.

The above list showcases the most widely known and used frameworks and standards. There are many more that might apply to a specific industry.

It is important to note that these frameworks and standards require substantial interpretation before implementation.

There are plenty of commercial organizations available to support organizations in framework selection, deployment, and monitoring compliance. They also provide qualification and training courses for your employees for the use of such frameworks and updates and their implications.

Frameworks and standards are supporting tools and should be looked at and used for such purposes.

A Word of Caution

While frameworks and standards provide a lot of benefits, they do not give you everything and they must be used in the right way and by practicing professionals who are qualified and competent.

For example, on its own, a framework will not provide you with a strategy, the skills needed, or an execution plan. These must be generated by the organization itself. Rather, the framework will help position this in the right direction and provide guidelines on the components needed.

Use of any standards should also be limited to the extent needed and fit for purpose so that it does not become an overhead.

Certifications are important but need to be acquired to serve a purpose not just for the sake of being certified. Going after certifications against too many standards will prove to be quite expensive.

Other Frameworks

While the above are well-known standards and frameworks, there are other projects or topic-specific frameworks that are presented quite often by major consultancy firms.

For example, McKinsey has presented a framework to come up with an effective digital transformation plan as it covers the full lifecycle of the project. This model depicts the current stage, starting point, the desired stage the organization aims to reach, and the journey in between, along with needed resources. (https://www.smartinsights.com/manage-digital-transformation/digital-transformation-strategy/structure-effective-digital-transformation-plan/amp/)

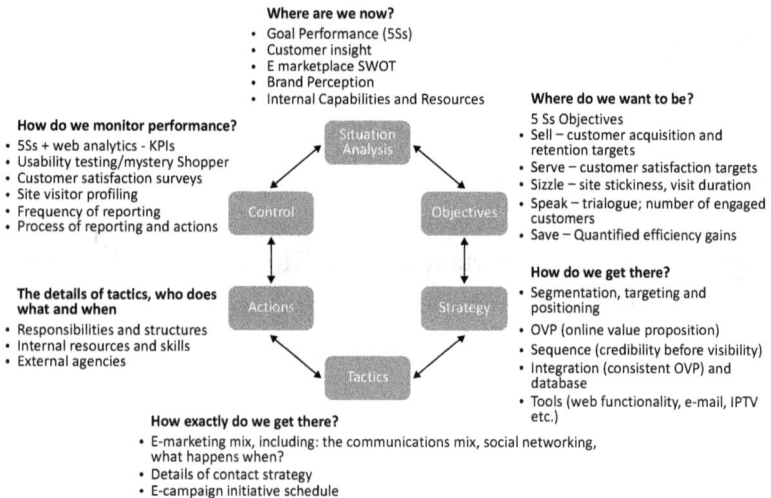

Where are we now?
- Goal Performance (5Ss)
- Customer insight
- E marketplace SWOT
- Brand Perception
- Internal Capabilities and Resources

How do we monitor performance?
- 5Ss + web analytics - KPIs
- Usability testing/mystery Shopper
- Customer satisfaction surveys
- Site visitor profiling
- Frequency of reporting
- Process of reporting and actions

Where do we want to be?
5 Ss Objectives
- Sell – customer acquisition and retention targets
- Serve – customer satisfaction targets
- Sizzle – site stickiness, visit duration
- Speak – trialogue; number of engaged customers
- Save – Quantified efficiency gains

How do we get there?
- Segmentation, targeting and positioning
- OVP (online value proposition)
- Sequence (credibility before visibility)
- Integration (consistent OVP) and database
- Tools (web functionality, e-mail, IPTV etc.)

The details of tactics, who does what and when
- Responsibilities and structures
- Internal resources and skills
- External agencies

Situation Analysis — Objectives — Strategy — Tactics — Actions — Control

How exactly do we get there?
- E-marketing mix, including: the communications mix, social networking, what happens when?
- Details of contact strategy
- E-campaign initiative schedule

Processes, Procedures, and Policies

Processes, procedures, and policies form the base of the building block of the operational model in the organization.

They are one of three pillars of the digital strategy (the other two being people and technology). They define how the organization's operations function in a structured and integrated way, and they form the basis of the organization's knowledge base.

Most successful organizations differentiate themselves in terms of how they define their processes, policies, and procedures.

To achieve success, you must have effective and efficient processes that are properly integrated, covered with detailed procedures, and governed with adequate and clear policies.

Many standards evaluate and mandate the existence of some processes, procedures, and policies—for example, security incident management processes, disaster recovery procedures, and email policy.

They do not define the process, policy, or procedure for the organization. This is left for the organization to define as it sees fit. However, they require their existence and effectiveness.

These can be measured by looking at results and through audits.

Here are simple definitions to clarify the difference between process, procedure, and policy:

- A *process* can be defined as a series of tasks that lead to a certain desired output.

- A *procedure* is how a certain task is carried out. It provides detailed instructions on how to execute the task.
- A *policy* is a clear description of a set of rules that describe acceptable and expected methods, behaviors, usage, etc. for a particular function, system, or tool in the organization. Policies are the cornerstone for good governance.

Benefits of Clear Processes, Procedures, and Policies

The existence of the right processes, procedures, and policies can generate many benefits:

- Defines and establishes the desired level of quality, effectiveness, and efficiency
- Provide a secure and lawful way of doing things.
- Ensure consistency
- Build an integrated block of control
- Outputs at different stages can be measured and hence provides the basis for continuous improvements. Those form a major part of your operational KPIs (Key Performance Indicators). We will talk about KPIs in detail in the next chapter.

It is important to note that operational effectiveness cannot be achieved unless the process effectiveness and efficiencies are continuously measured and improved. They are not static.

Procedures need to be reviewed regularly to ensure that they are up to date. Policies need to be adequate and must strike the right balance between control, risk, and flexibility to achieve the best results.

In summary, the combination of having effective and efficient processes, clear and balanced policies, and simple and clear procedures form the key to successful execution.

5

PERFORMANCE MANAGEMENT

To achieve your desired end state: Define it, measure it and control it

"When digital transformation is done right, it's like a caterpillar turning into a butterfly. But when done wrong, all you have is a fast caterpillar"

— George Westerman, MIT Sloan

Imagine a ship undertakes a voyage and the captain does not bother to pay attention to the speed and direction of the ship and has no visibility on the state of the ship. This may result in the ship never reaching its set destination – or even any destination at all!

Similar is the call for a digital transformation program – in the absence of an effective performance management system, the project may never successfully reach its destination.

By definition, performance management is to continuously monitor and ensure that individuals, teams, and the overall organization know what they should be doing, how they should be doing, and how they are doing within a clear and well-understood framework.

To ensure the successful execution of any project, a mechanism must exist by which you can measure and see how you are performing during the project phase and post-project phase.

Defining, measuring, and controlling performance is one of the key elements of a successful governance model. This includes all phases of digital transformation journey without exception.

It is important to note that you need to monitor relevant parameters and what matters—and avoid measuring too many indicators or unimportant ones. Doing so can lead to confusion and even wrong decision-making.

Visualize a scenario whereby we are driving to a place that is 150 km away. We have a set goal to cover this journey in 90 minutes.

What are the relevant and crucial factors in the assignment which should be kept under consideration and continuously monitored for the duration of the journey until we reach our target destination?

The relevant factors (we call them Key Performance Indicators) are:

1. Having enough fuel in the car
2. Good battery condition
3. Right tire pressure and conditions
4. Speed of the car
5. Road and weather conditions
6. Expected traffic

We need to define these and monitor them continuously so that we reach our destination in time and safely. The basic concepts in any project including digital transformation are similar, even though the situation may seem more complex, costlier, or have a longer duration.

The parameters to be monitored (called Key Performance Indicators, or KPIs) in a digital transformation journey may be relevant at various stages of the project.

The KPIs may belong to business segments pre-project, during the project, and post-project.

The business segment-related KPIs deal with expected business improvements (for example processes enhancements, financial benefits, productivity improvements, and corporate image).

The project-related KPIs deal with project progress and project governance-related aspects.

We introduce a new concept here, which we have successfully tried out in many projects of taking a "snapshot" view of the current state of the business-related KPIs and keeping that as baseline or reference to be measured against post-project goals and targets.

For example, if we define a KPI called 'expected benefits' and set a target of USD $1 Million in two years, we can have a snapshot of the actual benefit every quarter to see if we are on the right track. This new value can be kept as a reference point for future comparison.

All such business-related KPIs and snapshots must be determined together with the business.

There are different models available for managing performance and each organization uses what suits them best.

The challenge is not in the model itself but in selecting and using the model correctly. From our experience, we have found that the biggest challenge facing organizations is knowing what to measure, when, and how to measure it.

Measuring the wrong thing could lead to taking wrong decisions and hence deviating from achieving the main objectives.

Without having clear visibility over performance, one will not be able to achieve the desired targets.

Therefore, it is important to define key performance indicators that are relevant, meaningful, accurate, and useful. In addition, such KPIs must be shared across the organization and must be clearly understood by everyone – especially the larger project team and decision-makers.

What Are Key Performance Indicators, and Why Are They Important?

A key performance indicator is a quantifiable measurement of an important parameter that is a 'must have' for successfully achieving a target or goal.

The word 'key' is particularly important. These indicators should be defined clearly and must be measured in the right way. Defining the wrong KPIs or taking wrong measurements will lead to making wrong decisions.

The importance of KPIs comes from the fact that they provide insight and information that can be used to address any issues, concerns, and risks as they arise. Knowing such KPIs at an early stage enables decision-making and corrective actions to ensure achieving objectives.

Monitoring and reporting certain parameters makes it possible to become aware of concerns before they become difficult issues and problems. This will save time and money.

What Constitutes a KPI That Matters?

Defining a KPI that is important and serves the purpose is not an easy task. We recommend the following guidelines in selecting KPIs:

- A KPI must be a key success factor whereby the absence of achieving the performance of this parameter means that the success of the project or achievement of the goal or objective is in jeopardy.
- A KPI must be relevant to the project or the objective. We have come across cases where some KPIs used are quite generic and are not related to what is being sought.
  This happens quite often when mixing between benchmarks and best practices on one side, and project

performance or a specific organizational objective on the other side.

For example, comparing a digital transformation project with other projects might not be right. Every project has its own specific challenges and KPIs used must be relevant to that specific project.

- A KPI must be measurable. If you cannot measure something, you will not be able to control it.
- The KPI must be measured correctly and timely so that it reflects the real value of that KPI. Unfortunately, many times the right KPIs are defined but they are not measured correctly or in a timely matter. For example, many organizations measure turnover as a key measure for the retention of employees. This makes a lot of sense, but in case the organization is going through a downsizing phase and has intentionally reduced manpower, then the number of employees that have been let go should not be counted toward turnover ratio. Instead, only voluntary resignations should be considered.
- Any KPI must be meaningful and well understood by all those concerned. There is no point in defining a KPI that cannot be understood. This might sound strange, but it is quite common that KPIs are defined that have been taken from a certain standard or industry and blindly used. When asked about what value or purpose this KPI provides, the answer often is: it is an industry-standard!!

Broad Classification of KPIs

Large projects such as digital transformations require a number of KPIs. However, those KPIs are used by different people in the organization.

Therefore, it is useful to classify the KPIs so that they become clearer and used by the right people in the organization and in the right context.

Here is a list of some possible KPIs:

- Strategic KPIs
- Operational KPIs
- Financial KPIs
- Internal KPIs
- External KPIs
- Leading KPIs
- Lagging KPIs
- Project KPIs
- Functional KPIs
- Process-related KPIs

KPIs also can cover certain areas within a function— like data-related KPIs, service-related KPIs, and security-related KPIs.

We will limit our discussions to those KPIs that are of importance to CEOs.

Strategic vs. Operational KPIs: As the name indicates, **strategic KPIs** are those that have a strategic impact on the organization. For example, market share could be considered a strategic KPI.

Those are the KPIs that need direct attention of the CEO and have a long-term effect on the organization. Those are also more difficult to define and measure.

While market share is usually easy to measure, most strategic KPIs are difficult to measure and in many cases, subjective judgments are used to measure them. A good example would be measuring change readiness or employee morale within the organization.

Operational KPIs are those that are used to measure operational performance—for example internal service performance, system response, or budget performance.

Operational KPIs are also important – and without knowing how well you are operating; it is difficult to improve.

Which KPIs are strategic, and which are operational depends on each organization. For a CEO, a dashboard should exist that includes and reports a set of strategic and operational KPIs on regular basis.

Leading vs. lagging KPIs: Leading KPIs are those that provide you with information that can be used in forecasting and predicting the future through trend analysis. It can indicate to you where you are heading.

They are very important as they will enable you to take action early to correct the situation if you need to. For example, measuring the rate of increase in storage usage, allows you to increase the storage capacity before you run of storage.

Lagging indicators, on the other hand, provide you with historical data on how you have performed in the past.

They are also useful for performance improvements. Without knowing how you have performed; you will not be able to improve as you will not have a reference point.

While both types are important, unfortunately, most companies miss the use of leading indicators and use only lagging indicators. Doing so can lead to issues related to agility and scalability. That does not mean that lagging indicators are not or less important. It only means that the complete picture is not being seen.

Means vs. Ends: Many things can be measured. One way to classify what is being measured is to see if it is an end or if it is a means to an end.

Unfortunately, this is one of the biggest areas of confusion in many organizations. While both are important, they should be used properly and for the right purpose.

A good example would be measuring the quality of a product. One KPI would be the number of defects, or the number of failures within a specific period once in use. This would give you an indication of how good the quality is. This would be considered an end. On the other hand, the number of quality checks could be a good measure that will lead to better quality (fewer defects and fewer failures).

Means and ends are quite like leading and lagging indicators. The important thing here is that they should be used adequately. In many cases, we have seen KPIs that measure means but are reported as ends and this leads to the wrong interpretation of the KPI.

Other KPIs: To ensure effective governance, some KPIs could cover project management, change management, people management, financial, security, data, and technology.

Each of those KPIs provides valuable input to the concerned people. It is worth mentioning that financial KPIs are vital, and we are sure that those will be present all the time. We have not gone into detail for those KPIs as the purpose is to focus on digital transformation-related ones.

Important Things To Consider

In addition to the above, there are some key aspects related to KPIs that need to be considered.

- **The ideal number of KPIs:**
  Ideally speaking at the organizational level (CEO), you should have between 6 to 10 KPIs, maximum.

  Those KPIs should be mostly strategic ones. Some operational KPIs can also be used. Other KPIs could be defined at the next level.

- **Tools to define and measure KPIs:**
  There are many tools that help in defining KPIs at the organizational and divisional levels. Using tools will help quite a lot in ensuring alignment across the organizations. They also provide guidelines, clear definitions, and a structured way to approach KPIs. In addition, those tools have been used widely and have proven their effectiveness provided they are used properly.

The use of such tools requires knowledge and proficiency, and you must train your people on using them. Many institutions provide training and even certifications on using those performance management tools.

One of the best tools, in our opinion, is the balanced scorecard. It is a strategy execution tool that provides you with a complete and integrated picture of how well you are executing your strategy and how you are performing. However, it is important to spend time and define the right inputs to the tool so that you get the desired output.

You can use other tools that are more suited for specific purposes such as project management in combination with the balanced scorecard (or any performance man-agement tool) but make sure that they are integrated and KPI definitions across multiple tools are unified.

- **Use of standards:**
  Many standards identify what KPIs to be used and how to measure them. These are usually specific purpose KPIs. For example, ISO27001 provides KPIs related to information security that can be used in the organization.

  Using such standards can help identify correct defini-tions and measurements. They also help to benchmark against other organizations. ITIL also provides many KPIs related to service delivery. In many cases, regulatory requirements define how those KPIs are measured.

Some Examples of Digital KPIs

In the new digital world, new KPIs will emerge and some of the existing KPIs can remain valid. What is important here is to ensure that the defined KPIs are the right ones and that they serve the purpose. Following are some examples of digital KPIs:

Strategic KPIs include digital investment return, customer experience and satisfaction, time to market of new product or service on the digital platform, talent, and skills capability.

Operational KPIs include system availability, incident and recovery rates, system utilization, and system stability.

Financial KPIs could be budget performance, process cost, net margin, and conversion cost.

Security KPIs examples include number of security incidents, number of data breaches, and how up do date the software and tools are.

Technology KPIs include scalability of the platform, Obsolescence, and openness of the systems and solutions.

Project-related KPIs include delivery, budget, quality, and scope variation.

While we have listed some examples of KPIs, the list is by no means complete. There are hundreds of potential KPIs. The key message is to measure what really matters and focus on those measurements.

From our experience, the one KPI that has been missed or did not get enough attention in most failure cases of digital transformation projects is **customer experience**.

As an example, in an implementation of a major Tier-1 ERP system a large business conglomerate having different lines of businesses, including large automobile distributorship of a huge international retail brand. This aspect was missed out as detailed below.

When the implementation was completed in the automobile distribution business, a massive customer queue built up in its parts distribution center, resulting in unreasonable delays in processing customers' orders, which led to customers' frustration and tempers ran high to the extent that some customers started behaving in a very irate manner with counter employees, complaining that they used to get the parts much faster previously.

This happened because the new solution was never examined from a customer interface perspective.

This issue was anticipated and addressed by conducting a simulation exercise involving many people from the organization, family members, friends, and others in the next implementation. During this simulation exercise, the real business transactions were carried out with workflow monitoring.

The implementation and solution adaptation was much faster after this. Even after switching on the new solution for commercial business transactions, businesses can continue to monitor the system performance and customer experience by becoming mystery shoppers.

This may happen for various reasons, such as not engaging customers in the process, wrong timing of introducing digital platforms, not providing training, preparing customers for change, or assuming customer acceptance of process change among many other factors.

Moving forward, in the next chapter we will address cyber security, a topic that has probably kept many CEOs and boards worried. With the increase in the use of technology, the exposure to cyber security risk increases as well.

In the next chapter, we will address this issue and present how to deal with cyber security from a CEO's perspective.

6

CYBER SECURITY

Protect your Intellectual Property, now and always

"It takes 20 years to build a reputation and a few minutes of a cyber incident to ruin it."

— STEPHANE NAPPO

Would anyone want to keep their front door open, thereby extending an open invitation to robbers? Of course not. Therefore, it is mandatory to keep all possible entry points always protected. In the digital world, this is equivalent to potential Cyber Security threats if vulnerabilities are not dealt within time.

Cyber Security is one of the most discussed and critical subjects in modern business environments. It is no longer just confined to corporate digital systems but has extended far beyond the world of business.

In this chapter, we will focus on cyber security in the context of digital transformation programs in organizations.

Have you ever thought about what may happen if your organization's classified information becomes available to undesired and unauthorized parties, or some bad elements penetrate your corporate digital systems and lock it down for a ransom? Or even if a disgruntled employee deletes or leaks important information from company systems? None of these scenarios are hypothetical – they happen frequently in the real world.

To ensure that our systems run smoothly and thereby allow us to achieve business continuity, we must protect all our information assets and deploy effective measures in place to avoid any business disruption.

You must have read in the media about numerous, almost daily systems hacking and other compromises leading to business disruption, revenue loss, and image tarnishing.

In the first quarter of 2020, 1.7 million phishing attempts were made in the United Arab Emirates alone. Some big scams were carried out involving billions of dollars, and an international financial exchange house was put out of business for a few days and was able to restore business only after paying a ransom to hackers.

Incidents such as these are many and require businesses to remain vigilant and proactive to safeguard their environments from any threats.

In a recent survey, 55% of companies said that security was the number-one challenge they face when implementing digital enablement technologies. (https://appinventiv.com/blog/appinventiv-digital-transformation-guide/amp/)

The chart below shows that cyber security threats top the key challenges in digital transformation projects.

(https://www.netsolutions.com/insights/challenges-to-a-successful-digital-transformation-and-how-to-overcome-them/#4-digital-security)

Cyber security has been identified as the top key challenge of digital transformation.

Perceived Challenges of Digital Transformation

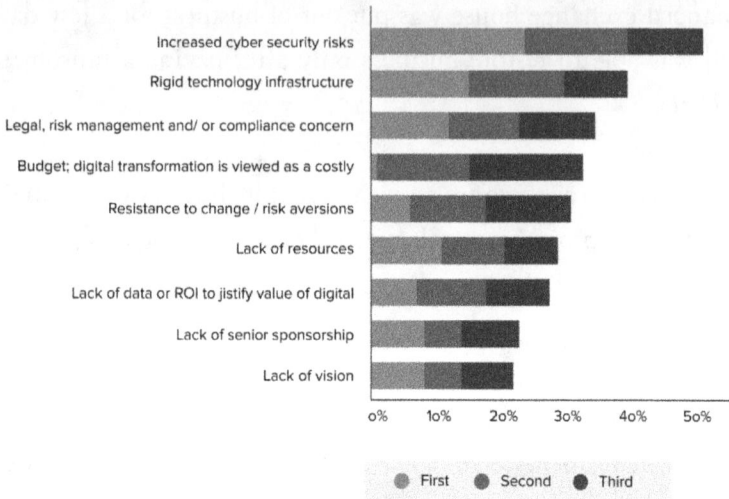

Challenge	Value
Increased cyber security risks	
Rigid technology infrastructure	
Legal, risk management and/ or compliance concern	
Budget; digital transformation is viewed as a costly	
Resistance to change / risk aversions	
Lack of resources	
Lack of data or ROI to jistify value of digital	
Lack of senior sponsorship	
Lack of vision	

0%   10%   20%   30%   40%   50%

● First   ● Second   ● Third

The compromise on cyber security could take place via one of the below or a combination of below factors

1. People
2. Systems
3. Technology

In this list, people are the most vulnerable component and must be given focused attention in terms of awareness, education, and disciplined behavior concerning acceptable usage of company systems, information, and assets. In other words, the people factor is usually the weakest link.

Basics of Cyber Security

Before we cover the preventive and remedial measures regarding cyber security, let us understand what cyber security is.

Cyber security deals with protecting organization assets and looks at ensuring the availability of systems and data, the integrity of data (ensuring data is intact and can be trusted), and confidentiality (ensuring that access to data and information is limited to those authorized)

Cyber security can be applied to software, data, network, mobile devices, and others. Every item that is connected to your system is part of the cyber security landscape.

The physical security of information and technology assets is also important and needs to be looked after.

There are many types of threats such as password attacks, identity theft, ransomware, insider threat, information and data leak, social engineering, email threats, malware viruses, and spyware attack, among others. This book is not a technical one. Therefore, we are not going to talk about the types of threats in detail. Many publications provide this information if the reader is interested.

Many approaches and frameworks can provide structured guidance to address cyber security.

We recommend the following framework as an effective way to approach and deal with cyber security threats. This approach has four phases: define, prevent, detect, respond and correct.

- **Define**

  This phase starts with defining the landscape of threats that the organization can be exposed to. This depends on the number and type of systems in place. It also depends on the depth (how far deep into the organizations processes do the IT systems and technologies cover?) and width (how many business units and processes across the organization and beyond (other parties such as suppliers and customers) is covered by the systems and solutions deployed?).

  It is essential to define the scope. Otherwise, the business may end up spending money and buying technological solutions to address threats that are irrelevant to the organization and that never existed.

- **Prevent**

  The best thing to do is try and prevent any incident. Although it is practically impossible to achieve 100% prevention, putting the right measures will prevent most of the common and known threats.

Many technologies provide you with such capabilities and you should use the right technologies. Unfortunately, many organizations overspend here and throw their money on buying far too many technologies and solutions.

Beyond a certain point, additional technologies do not provide any additional security. Instead, they may make your technology landscape complex and costly.

One of the key preventive measures is having clear and comprehensive IT policies, coupled with creating awareness among employees and other users of the system. Most breaches happen because of internal reasons: either due to unawareness or fragile IT policies.

- **Detect**
  Just because you cannot prevent all threats, you should not stay idle. Instead, go to the next level of dealing with cyber security threats by trying and detect a vulnerability or breach as soon as possible.

  By using tools, it is possible to detect breaches early. Many tools help in monitoring networks, performance, and data, and detect any abnormal patterns or behaviors. At this stage, it is possible to deal with the breach and eliminate or minimize the effect. Many intelligent solutions can help in this.

- **Respond and Correct**

  In case we reach a stage where a breach has occurred and was not prevented or detected, we will need to be able to respond quickly to minimize damage.

  Depending on the nature of the breach, different responses can be taken. What helps here is the availability of Disaster Recovery Plans (DRPs) and Business Continuity Plans (BCPs). We will talk about DRPs and BCPs later in the chapter.

  Following this, we will need to learn from this experience and put some corrective measures to avoid future occurrences of such breaches.

  It is important to note that, in case of breaches that affect outside parties, it is mandatory to inform all relevant and concerned parties, authorities, markets, and stakeholders in a timely and correct manner.

  The proper architectural design of networks, applications, and solutions is a key factor in addressing Cyber Security.

  In addition, using standards (such as ISO 27000 and NIST), staying abreast of the latest developments in cyber security, discipline in timely upgrades and patch deployments, and regular audits help a lot in reducing exposure as they provide a safer environment.

We may also consider being part of a reputed CSOC (Cyber Security Operation Control) to facilitate real-time 24X7 vigilance and monitoring of our systems. Many CSOCs also offer threat intelligence, which is part of prevention.

Best Practices Frameworks for Effective Cyber Security

For effective cyber security governance and implementation of tools and technologies to address cyber security, we recommend the following frameworks to be followed by the organization and digital transformation programs:

Cyber Security Policy

All organizations must have a cyber security policy that is clear, comprehensive, and well-thought.

This policy should be accepted and approved by the Chief Executive Officer of the organization (and/or the board of directors) and well-publicized and cascaded across all levels so that everyone in the organization becomes aware of the cyber security policy.

The policy should also be reflected in employees' onboarding contracts and performance appraisals. In addition, it also should be shared with all concerned parties outside the organization.

Education and awareness
There should be widespread, continuous, and effective programs to promote awareness about updates on cyber security threats and measures.

The organization should plan and conduct regular awareness sessions through all possible and relevant means. Mechanisms should be in place for the enforcement of security against cyber threats and how to guard against all such threats.

Nowadays, there are plenty of short video-based e-learning courses that create awareness, inform about the relevant dos and don'ts, and provide self-assessments to ensure understanding. We find these to be very effective channels of education.

Protection through technology
There are technological tools that help in addressing cyber security and should be used.

The most used platform for communication and information creation and maintenance comes with state-of-the-art protection tools. Similar features are also being made available by most of the software vendors with the primary aim of protection of data and intellectual property.

One example is multi-factor authentication, which requires dual verification of identity before giving any access to systems to the requester.

Real-time engagement of cyber security team

We recommend that the cyber security team is made an integral part of all digital transformation projects ecosystem, rather than being brought into the game at the end.

The cyber security team can guide the project teams in line with proper security practices and safeguards.

Once a system is developed or procured and configured, it must be validated by the cyber security team before any such system is moved into a production environment and before giving access to the system to any user.

Cyber security will check all such systems to find out if there are any vulnerabilities and whether the system can be compromised (hacked). In the case that any such weakness is identified, it should be fixed, the system should be revalidated and then deployed into production.

The cyber security team should be continuously engaged throughout the process of procurement, right from the RFP to vendor and solution evaluation and trial.

In many organizations, the security team will include a cybersecurity-related questionnaire with a Request For Proposal for the vendors to respond. The response of such questionnaires provides an initial impression in terms of the security ruggedness of solutions, vendors, or services.

The typical segments of such questionnaires are data protection, vulnerability management, identity & access management, physical security, application security, incident response, privacy, business continuity & disaster recovery, end of service support, etc.

Vulnerability Assessment and Penetration Testing (VAPT)
At regular intervals, organizations must carry out an exercise called Vulnerability Assessment and Penetration Testing (VAPT).

This is usually done with the help of qualified external agencies under strict non-disclosure agreements and guidelines.

Through this exercise, such agencies make attempts to penetrate the company network and come out with their observations regarding existing vulnerabilities in the technical landscape of the organizations.

These vulnerabilities are categorized under critical, high, and medium. The critical and high vulnerabilities must be addressed and fixed by the organization on a very high priority. Such exercise must be repeated at least once a year, preferably with different agencies year after year.

Simulated Phishing Attacks

To assess the organizational awareness and alertness on cyber threats, an organization should carry out simulated phishing attacks in the organization at regular intervals.

Under such an exercise, genuine-looking emails are sent to employees seeking sensitive information – like their company credentials – and results are analyzed.

We have seen that many initial attempts can "capture" employees who unknowingly share their credentials and sensitive information. As the exercises are repeated, that count starts to reduce, indicating that awareness is rising.

The results of such exercises at the number or percentage level should be published, but individuals' names must not be published. Rather, they should be contacted and communicated individually.

Key Recommendations

With the evolving technological landscape and changing corporate and climate situations, the cyber threats landscape and profile keep changing and evolving. This necessitates a very agile approach to cyber security from an organization's perspective also.

The consequences of a successful cyber-attack could be disastrous, and organizations must protect themselves by remaining one step ahead of the bad actors. It is therefore important to be aware of the ever-evolving new technologies because new technologies keep revising the threat landscape.

Considering the above, we recommend always keeping the following in mind:

Cyber security is everyone's responsibility

Gone are the days when organizations would think that cyber security is a technical issue and the responsibility lies only with a technical team.

In today's world, all of us own and share this responsibility. We must protect both ourselves and our organization. Therefore, everyone should play a very active role in cyber protection.

Cyber security is a dedicated function

Organizational cyber security should be handled by a dedicated core team and not with shared responsibility. This cyber security team may work closely with business continuity, disaster recovery, and audit teams.

This team should be independent of a conventional Information Technology team but work collaboratively because only collaborative working can give protection. The cyber security team should report to the Chief Executive Officer of the organization.

Standards in cyber security

There are various standards in cyber security. The most established and known is ISO 27001. It is important to remain aligned with established standards. While certification is not mandatory, it provides a higher level of assurance.

However, certain factors like the type of industry or geopolitical considerations may enforce strict compliance. Most regulations draw on established security standards like ISO 27001 and NIST and require compliance.

Some standards pertain to certain industries such as HIPPA (Health Insurance Portability and Accountability), HITECH, PCI-DSS (Payment Card Industry Data Security Standards), and GDPR (General Data Protection Regulation) among others.

Disaster Recovery and Business Continuity Plans

As stated earlier, it is important to have a disaster recovery plan (DRP) and a business continuity plan (BCM) in place. A disaster recovery plan, as the name suggests, is a formally defined document that provides instructions on what to do in the case of a major incident.

The purpose of this is that when an incident occurs, you don't need to spend time trying to define what should be done. Instead, the plan is already in place, and you should execute it.

Of course, such plans should provide the path to the most effective and efficient way of recovery. That is where the term 'plan' comes into place.

This book is not about disaster recovery planning, but it is important to know that having such a plan in place requires lots of thinking and arrangements. It touches on your data and backup and restores policies and procedures, having remote sites as backup, and quick identifications of key systems and data.

What is important to note here is that unfortunately, most companies do not have an effective disaster recovery in place.

A business continuity plan, on the other hand, refers to how the organization will continue operating in the absence of the system (until the system is recovered).

It involves the identification of key processes, systems, and data that is made available to the organization that will enable it to continue operating whilst the systems are down.

Obviously, the organization will not be able to operate with the same effectiveness and efficiency in case of the absence of the systems in comparison to the presence of all systems. Therefore, the organization needs to identify which processes are top priority and need to be running, and arrangements should be made to enable those as quickly as possible.

To ensure the effectiveness of both disaster recovery and business continuity plans, regular drills should be conducted.

Final word:

Always remember: if a vulnerability exists, it will be exploited.

The digital world is about technology, and digital transformation is powered by technology. As technology enhances, it enriches the digital experience. Therefore, it is important to identify the new emerging technologies that could shape our experience in the future. The next chapter will focus on these emerging technologies.

7

EMERGING TECHNOLOGIES

Hold-on to remain on course – don't get swayed

"To master a new technology, you have to play with it."

— JORDAN PETERSON

Civilization is going through rapid technological developments, affecting all walks of life. We are in an age of massive scientific rediscoveries. No sooner than a digital transformation program is initiated, the related technological ecosystem starts redefining itself, raising an ongoing question about whether we continue with our digital transformation with the original technology landscape or realign with the newer technology and versions. It's a good problem to have, with plenty of implications. Projects are influenced by emerging (or new) technologies. How should these emergencies be treated? Should we move into ostrich syndrome or take note of it and align intelligently?

This chapter will enable readers to get a snapshot of emerging technologies. Snapshots of emerging technology will soon become outdated and therefore the validity of such information is very short-lived. Therefore, it is important to establish a strategy to deal with technology. Do you want to be a technology chaser or an adopter of mature technologies? Each approach has its advantages and disadvantages.

With the rapid pace at which new technologies are emerging, it is difficult and almost impossible to cope with them. The challenge is that if you ignore emerging technologies, you might fall behind, which may have a business impact. On the other hand, if you continuously try to adopt technologies, you are faced with two issues:

1. Keeping up to date with emerging technologies and selecting the right ones
2. Selecting a technology too early in its life cycle may lead to a huge financial burden and risk

All technologies have a maturity cycle: some of them mature quickly, and some may never mature. To make the best use of emerging technologies, it is important to be able to assess the value it brings to the organization. Being a first-mover has a huge advantage, but it also poses a lot of risks. It is important to be able to select the right technology at the right time and strike a balance between a first mover and maturity. Agility plays a major role here.

'Governing Digital Transformation and Emerging Technologies: A Practical Guide', a joint report from the National Association of Corporate Directors (NACD) and Marsh & McLennan Insights, offers a new roadmap for successful corporate governance in the digital age. This report draws from primary research conducted through interviews with company directors as well as a survey of 200 NACD members and outlines five foundational principles to help prepare directors for navigating the complexities of digital transformation and emerging technologies:

1. Approach emerging technologies as a strategic imperative, not just an operational issue
2. Develop collective, continuous technology-specific learning and development goals
3. Align board structure and composition to reflect the growing significance of technology as a driver of both growth and risk

4. Demand frequent and forward-looking reporting on technology-related initiatives; and

5. Periodically assess the organization's leadership, talent, and culture-readiness for technological change

Each principle includes specific recommendations and questions for consideration to assist boards with charting potential.

Evolution of Technology and Emerging Technologies

The current corporate world and civilization at large are now approximately into the 50th year of the digital age. The 1980s were the era of low-capacity huge size mainframes, 3rd generation programming languages, and secured data centers. The era of the 1990s may be termed as an era of re-engineering when the industry started revisiting and redesigning the basics and architecture of digital frameworks. Thereafter, the digital age moved into the velocity drive and that is continuing. This period of the digital era was busy with architectural changes and design re-engineering. In the meantime, the arrival of the Internet brought in a massive transformation in the digital era, the way business used to be conducted, and the visible impact of society.

Major players like IBM, Microsoft, Oracle, Dell, HP, and Intel established themselves in the industry. These players are in fierce competition with each other and in an aggressive race to capture more and more market share – and in the process, proprietary technologies started dominating the industry. It was an era of various "camps" where organizations started becoming the customers of exclusivity. To support and promote such culture and psychology, many independent analyst firms came into existence and their business boomed. All these are at the cost of customers. Thereafter, these major players started realizing that partial co-existence may be a better option for them and the industry also. This approach of co-existence with fierce competition caused a massive explosion in business volume for IT vendors and the extent of computerization. This expansion created several services delivery companies across the globe.

A major event happened closer to the year 2000 known as Y2K, which had the serious potential of disrupting the corporate world, services industry, utilities, and medical industry – wherever digital systems were in use. The problem was that most of the digital systems that existed at the time were using 2 digits to represent the year (for example, 98 meant 1998), and as the year 2000 was close to arrive, the year number would change to 00 and this could cause the system to halt. This problem had to be addressed before the arrival of the year 2000. This challenge turned out to be a massive opportunity for countries like India, which had a large population of young engineers who are familiar with the English language. Huge demand was created in the USA and Europe, and Indian IT Services Company got a big boost and captured the IT services segment of the business, becoming the extended arm of the major IT players like IBM Oracle, etc.

Then in the last decade, as the internet became more mature and rugged, the Digital industry started moving towards the cloud. This wave of the cloud journey also started on a captive and proprietary note. But, then another set of players entered the industry and their business was to provide cloud platforms, irrespective of which major IT vendor's systems the organization needed to move to the cloud. Indeed, in the last decade, a lot of development and research has gone into cloud developments and players like Amazon Web Services, Google Cloud Microsoft Azure, etc. started making their presence felt.

As Bill Gates has mentioned and predicted in his book, Business @ the Speed of Thought, more than two decades back about the digital nervous system, which talks about information flow and information availability, it is happening the same way.

Now we are in an era of Artificial Intelligence, machine learning, and agile ways of project delivery and development. The monolithic dominating ERPs are being challenged by fit-for-purpose pointed solutions. System development parameters are going through a paradigm shift and this trend is giving rise to another set of industry players. It is an interesting time of shift of power amongst the power brokers and we from the business side need to tread very carefully to avoid any professional trap.

Among the many emerging technologies, some have reached a level of maturity that can be considered by organizations, provided a proper assessment and feasibility is done and the technology is applied properly. Such technologies could be either general (those that can apply to all industries), or industry-specific (i.e., Healthcare, education, financial.). The following is a list of some of the general technologies:

- Cloud technologies /Distributed Cloud
- Internet of Things (IoT)
- Mobility
- Dev-ops
- Robotic process automation (RPA)
- Artificial intelligence (AI)
- Blockchain
- Face recognition
- Augmented Reality/Virtual Reality/Mixed Reality
- Human augmentation
- Hyper automation
- Microservices
- Managed and intelligent ERP

In its report on Strategic Technology Trends 2021, Gartner has highlighted seven emerging technology trends. It is worth noting that some of these trends have ethical and societal implications. One trend which stands out is the Internet of Behaviors (IoB). Other major emerging technology trends are

Distributed Cloud, intelligent business, hyper-automation, and AI engineering.

How to manage new technologies?

Just because there is a new technology that everybody is talking about or has started using, it does not mean that you should procure it and use it in your organization. We recommend that you follow the following steps when dealing with new technologies:

- Perform a feasibility study to arrive at the value that such technology will bring to the organization. Here you are looking mainly at the financial benefits that such technology can bring to the organization
- Ensure that the selected technology aligns with the overall business strategy, direction, and objectives
- Ensure that the new technology integrates well with your existing technology landscape and business processes
- Assess your capability to implement such technology (Internal resources, skills, and knowledge)
- Identify the impact that this technology will have on the whole echo system (especially customers and employees), and accordingly address change management requirements
- Ensure a detailed and realistic project plan is in place and that adequate governance exists that addresses financial, risk, and project issues

New Technology Challenges

Introducing new technologies also bring in many challenges along with it, as listed below:

- Business Challenges:
  - Cost
  - Alignment with the company's overall vision and mission
  - Alignment of processes with new technologies
  - Introducing new risks
- Technology Challenges:
  - Integration and compatibility with existing systems
  - Meeting existing standards and open systems (i.e., propriety technology)
  - Finding the right and capable implementation partner
  - Bringing technology for technology's sake (Unfortunately, many CIOs today opt for getting new technologies just to be seen as new technology adaptors
- Skills Challenges:
  - New technologies will require new skills for employees, outsource parties, customers, etc.
- Change Management:
  - New technologies bring about change at all levels. With multiple new technologies, change management can prove to be one of the most difficult challenges to deal with

An organization during SAP ERP implementation went through a change of technology change decision. It was a large group with businesses in numerous industry verticals including Automotive, Retail, and Logistics. The project was started with SAP / R3 IS-Retail, but just after the business blueprint got over and the project was about to enter configuration and conference room pilot, the discussion started that the project should change its direction and go for SAP R3 IS Automotive.

Getting overwhelmed by new technologies might turn out to be counterproductive if adopted without adequate assessment and readiness.

Each new technology does not come with a promise of better outcomes and results. You may not want to buy a car with a turbo engine if your requirement is a normal drive on city roads rather than on a racetrack.

8

TALENT MANAGEMENT

Acquire, develop, and keep the right talent

"Talent hits a target no one else can hit.
Genius hits a target no one else can see."

— ARTHUR SCHOPENHAUER

All of us can buy the best available tools in the market but most of us cannot fix common and relatively simple problems at home. We ask the help of technicians, plumbers, and the like to come and fix such problems. Tools do help but without the right skill and talent, it is almost impossible to get the job done right.

Similarly, when it comes to digital transformation, organizations can buy the latest and best technologies, but it is the human factor that makes things happen. Without having and maintaining the right set of skills, the digital transformation journey can become quite cumbersome.

The ability of an organization to maximize the use of new technologies to generate value for the organization is dependent on the organization's capability in terms of the talent and skills it owns.

Why is Talent Important in Digital Transformation?

As new technologies emerge at a fast pace, access to skills related to the new technology becomes limited. Only talented employees can learn such new technologies fast enough to be able to drive digital transformation.

Organizations are spending millions of dollars on technologies but miss investing enough in acquiring and developing necessary talent. The result is lost opportunities in terms of getting the best value from such investments.

In a recent survey of CIOs, 69% of them said that they are re-evaluating their skill sets due to shifting priorities driven by digital transformation. (https://www.ey.com/en_us/cio/ how-digital-transformation-is-driving-talent-and-culture)

According to another study, having the right employees provides the organization with the potential to accelerate digital transformation by 20 to 30 percent. (https://www.pluralsight.com/blog/career/digital-dna)

Many believe that digital transformation is about people and not technology as the main drivers. This comes in two ways: change management on one side and talent and skills on the other side.

In the absence of some key skills, the whole journey could be in jeopardy. Some believe that we have entered an era of "wars for talent", whereby companies fight to acquire new talents that are scarce due to new technologies and a large number of digital transformation initiatives.

What type of talent and skills do you need for your digital transformation journey and how do you go about acquiring and maintaining them?

What skill sets are needed:

While the first thing that comes to mind is technological skills, the plain truth is that this is not the only type of skill sets you need. There are many types of skills needed:

1. **Technological skills**

   As you introduce new technologies, you will need people who are proficient in those technologies. How to best configure, use, and troubleshoot them.

   One might say that this can be outsourced. The answer is that maintaining the technology can be outsourced but using the technology in the best possible way to add value to your business require skills that combine the business part and the technology part.

   This is the main challenge. Finding people who have combined skills in business and technology is difficult. From our experience, this is true even at the highest level in information technology. Most Chief Information Officers lack the business part and are focused more on technology.

2. **Process engineering skills**

   Digital transformation requires reengineering and redesigning business processes. This requires knowledge of process design, integration, effectiveness, and efficiency.

   In addition, it requires knowledge of regulations, boundaries, data, reporting, and outcomes requirements of all stakeholders of the process. For example, when redefining a process in procurement, the requirements of accounts payables need to be considered.

To ensure a proper definition of processes, detailed knowledge of all areas of business is required.

3. **Change management**

   We have talked about change management and the role it plays in digital transformation. To succeed in change management, you need a team that has soft skills that can deal with people in the organization so that the required change takes place.

   Those are people who can deal with all levels in the organization from top leadership to casual employees. They connect with people and know the right way to approach people at the right time, pass the right message, use the right language, and assess the readiness of the individuals and the organization.

   Those people also require an understanding of the business, the amount and impact of change coming to the organization, and the consequences for the organization of not being able to change. They must be credible to be effective in their roles.

4. **Data management**

   Any change in process or system requires a change in the definition of data in terms of how this data is defined, calculated, captured, stored, used, and retained.

Many challenges arise such as: what about previous data? How to convert them? What about unstructured data? How about the relationship between different data sets? What about the new data sets that have come about because of the newly redefined processes?

Those are only some questions about data. Data is a broad topic. However, we will touch on some aspects of data governance in a separate chapter.

What is obvious is that there is a need to have such capability and skillsets within the organization to deal with data in the context of digital transformation.

5. **Analytical capability**

   Moving to the digital world, redefining the processes, and introducing new technologies will bring a world of data and information. To be successful, the organization must be able to analyze the data and make meaningful use of it in terms of decision making.

   Doing so requires people who have analytical capabilities, understand the business and the industry well, know the processes well, know the vision and objectives of the organization, and know the definition of data within the organization.

There are new roles that have come up in this regard such as data scientists, who can define an analytical framework that provides insights and intelligence derived from your data.

It is important to note that the above types of skill sets do not have to be in one person, but you can have teams that work in harmony and complement each other's skills.

Another important point is that the above types should work together and not in silos. By working together, desired results will be achieved; otherwise, there will be lots of gaps and misalignment.

Ensuring You Have the Right Skill Sets

A digital transformation journey requires staying up to date with development in technological solutions. An important part is to be able to cope with these developments by always having the right skills. We recommend doing the following to ensure that:

Develop an action plan to build the skill sets required

The first step is to identify the new skills required as you move on in your digital transformation journey. Next, assess your organization's current skill sets.

This will provide you with knowledge about existing gaps. The next step is to define an action plan on how you are going to fill those gaps, either through developing some of the skills or through hiring new talents.

This can be done at an early stage which provides you with time to fill the gaps. You can always borrow some talent from certain companies until you build your own talent pool.

Upgrade the skills of your employees

This refers to both those who are responsible for the day-to-day technology management as well as other employees.

In fact, the focus should be on other employees. They do not need to know everything about technology, but they need to know enough about the technological solutions deployed in terms of using them, what they provide, and so on.

Try and build and culture of learning within the organization. In other words, you need to create a workforce that is digitally aware and understands what digital solutions do, how they work, with the willingness to learn.

Hire or develop data analysis capability

Going digital means having quite a lot of data. To be able to get insights from your data, you need people with data analysis skills who can analyze data. This will provide you with visibility over performance, trends, and gaps which is essential for timely decision making.

Ensure top-level executives are digitally aware

Top leadership teams play a major role in creating a digital culture. They should take ownership, commit, and set good examples by learning the digital world, especially the industry in which the organization operates.

This is essential to drive the digital transformation program forward. In the absence of leadership commitment, change will become difficult.

Develop the soft skills of your people

To succeed in change management, a team that possesses key soft skills such as communication, presentation, creativity, seeing the big picture, innovative, positive mindset.

Most of these skills can be developed through training. It is important to ensure that those who are in key positions have such skills.

Review your policies

The new reality of digital transformation and the fight for talent requires that you review some of your human capital policies, especially those related to recruitment, training, and retention of employees. You need to have more flexible policies to allow for the demand.

With so much of the expenditures going to the acquisition of technology and solutions, it is essential to invest in retaining talent and in continuously upgrading the skills of the workforce.

Unfortunately, from our experience, most organizations do not pay enough attention to investments in their own employees to support the digital transformation journey.

On the contrary: we have seen some organizations invest in training some of their customers so that they will have a seamless transition to the new digital world. Those are the visionary ones who have can see the future and bet on succeeding.

9

DATA GOVERNANCE

Properly governed data fuels business success

"Governance allows organizations to use critical data to drive the organization"

— DICK TAYLOR

If you were a CEO of a company and asked your CFO and your Chief Marketing Officer to provide you with a particular customer's business over the last three years, and you received different figures, would you be happy? Which one should you believe?

This is a common problem, and it is due to inconsistencies in data. Data governance addresses all issues to do with data.

Data serves as the backbone of all information systems. No system works without data. Every transaction or interaction requires data as input and generates data as output.

As more and more digital solutions are deployed, more data is generated. Over the last few years, the amount of data generated has been continuously rising. While in the old days, data was simple characters and numbers, today data comes in many forms (i.e., pictures, images, videos, web activities).

This data can be used to generate reports, perform analyses, evaluate performance, and other business purposes.

However, it is important to ensure that all those data are defined and used properly, by the right people, in the right form, and are protected from misuse. In other words, data must be governed properly.

What is Data Governance?

Data is one of the most important assets that organizations have. Data governance consists of policies, procedures, and guidelines that control how data is managed. Effective data governance ensures that data is consistent and trustworthy and does not get misused.

The main purpose of data governance is to ensure the integrity and security of data are maintained at the time. In addition, it ensures that usage and manipulation of data are done properly and in line with all standards and regulations. It also provides a ground for audits through proper documentation.

In other words, data governance means that data within the organization is managed and controlled through all phases of the data lifecycle

There are several definitions of data governance. The Data Governance Institute defines data governance as a "practical and actionable framework to help data stakeholders across any organization identify and meet their information needs."

A simpler definition that we particularly like states that data governance is the process of managing the availability, usability, integrity, and security of the data in enterprise systems, based on internal data standards and policies that also control data usage.

(https://searchdatamanagement.techtarget.com/definition/data)

The importance of data governance has risen recently as increasingly digital solutions are put in place that generate large amounts of data that need to be collected, protected, and analyzed. Here are some main benefits that data governance provides:

- It provides top management with a high-level view of an organization's performance powered by clean data that can be trusted and a single reference for definitions of key metrics.
- It creates one version of the truth. This means that everyone in the organization is on one page when it comes to key parameters and key performance indicators whether they are financial or operational. This, in turn, creates a lot of value and eliminates unnecessary discussions and disagreements.
- It provides an environment that ensures security, usability, availability, and integrity of data. From poor experience, data integrity problems exist widely among most organizations. This leads to problems and conflicts.
- Data governance helps in breaking down data silos created by different systems. This will support achieving strategic business objectives.
- It improves the quality of data which results in fewer errors in data. This will lead to smoother operations, clean input to analytics and intelligence, and hence better and timely decision making.

Effective data governance leads to better data analytics, which in turn leads to better decision-making and improved operations support.

(https://www.bmc.com/blogs/machine-learning-vs-predictive-analytics/)

- Provides an environment that helps in meeting regulatory compliance and data-related laws, such as privacy laws.

Objectives of Data Governance

Data governance, if implemented right, can provide the organization with huge benefits. The starting point to define, understand, and deliver the key objectives of data governance is as follows:

Achieve data integrity
Data integrity means that data can be trusted. This can be done through elimination or minimizing data errors. Correct data provides information that is accurate and when used as input to decision making, it would result in better decision making.

Ensure adequate access and use of data
Data can be classified in many ways. Some data might be considered sensitive and need to be restricted. The main objective of data governance should be to ensure that people have access to the information they need and that they use it for the right business purposes.

To do so, the organization needs to classify data and define clear policies, procedures, and guidelines. Any misuse should be considered a breach and should be investigated.

Comply with all relevant laws and regulations

A key objective of data governance is to adhere to all applicable local and international data laws and regulations such as data privacy laws. Doing business on the internet means that you will have customers from different parts of the world. Different countries have different laws.

Improve data security

An important aspect of data governance is to ensure that all data is secure from illegal and unauthorized access either internally or outside parties. It is also important to ensure that data is not leaked by employees, especially key confidential data.

Components of Data Governance

Depending on the purpose, objectives, and scope, each organization can define the landscape of its data governance. The following are the key components that, in our opinion, should be included in any data governance program:

- **Data dictionary**

  A data dictionary contains common definitions of the organization's key data so that it is used by everyone in the organization to mean as per the agreed definition. For example, how does the organization define employee turnover, unit cost, or inventory costing?

Of course, all definitions should adhere to applicable standards, whether they are financial or technical. Standards generally provide flexibility and options in those definitions. However, data governance ensures consistency.

In addition to that, a data dictionary provides a catalog of data that is indexed and can use a reference for the meaning and definition of each data.

- **Master data management**
  Master data management ensures that you have one reference for each item. For example, details of an employee or a customer must be defined in one system.

  Unfortunately, as most organizations use multiple systems, they end up having multiple master data files. This leads to inconsistencies among different systems. For example, details of the same customer could differ from one system to another. This problem is more common where multiple business units exist.

  Through data governance, you can still use multiple systems but only one system holds the master data and the other systems would get the information from that system.

- **Data Classification**
  Data classification means dividing data into distinct classes. Each class has its own properties. Each class of data can be used as defined in the policies, procedures, and guidelines.

  Each organization can define the classes as they see fit. There should not be too many classes, as it becomes difficult to distinguish between different classes. As a general guideline, four to six classes are ideal.

  We recommend using the following four classes of data:

  1. **Public:** Data that is defined as public means that it can be shared with anyone inside and outside the organization.
  2. **Internal:** Internal data should be used and shared with anyone with the organization including all business units.
  3. **Confidential or restricted:** Access to this type of data should be limited to a few people who need it to perform their duties as the organization sees fit. A good example would be employee details.
  4. **Secret:** This is the most sensitive class and, in this class, the information should not be shared by anyone. This usually happens when an organization is looking at some key project, acquisition, or similar.

Some organizations might use other definitions or add more classes such as geography, business unit, or project in consideration of the number of classes.

How to build a data governance program?

The best approach to build a data governance program includes the following:

- Ensure that the program is owned, supported, and driven by the top leadership team in the organization.
- Form a dedicated team in the organization responsible for the overall program from the implementation phase to the maintenance and running phase. Make sure this team has the capabilities and skills needed.
- Define the purpose, objectives, and scope of data governance programs. Include how are you going to measure the effectiveness of the program by defining some key success factors.
- Define a roadmap and a detailed implementation plan. We recommend taking a phased approach and breaking down the implementation plan into three or four phases. This will provide more focus on the program.
- Define your data governance framework, identify the standards you want to use and the desired components.
- Develop policies, procedures, and guidelines. This includes relevant security policies and applicable laws and regulations.

Challenges in Implementing Data Governance Programs

Implementing a data governance program is not an easy task at all. There are lots of challenges that must be overcome:

1. The first challenge is to create and articulate a business case. The main difficulty comes in how to quantify the benefits. One way is to break the program into smaller chunks and start with the higher priorities.
2. Another challenge is how to start considering digital transformation projects that want to move at a fast pace. This might be seen as an initiative that hinders project timelines.
3. The scope definition is another challenge. How much data do you include? What about legacy systems? If you include all legacy systems, the scope becomes unmanageable, and the cost will increase drastically.
4. While developing the data dictionary, agreeing on the definitions among all is quite a challenging task. The larger the organization, the more difficult it becomes.
5. A big challenge comes from analysis tools that provide the facility to redefine data and create new fields and so on. This poses a threat in terms of misuse of data. This might not be intentional, but it causes inconsistencies. To overcome this, solid policies and procedures must be defined and enforced.

6. Another challenge is training and keeping everyone aware and up to date with the latest policy changes due to regulations or new systems.

CONCLUSION

Safeguard your digital transformation investment with solid governance

With the rapid increase in corporate businesses today embarking on digital transformation journeys, the fast pace at which technology enhancements are taking place, and the large investments, failure is no option.

However, success is not easy, as the studies have shown us that over 70 percent of cases have failed. The challenges faced in digital transformation journeys are not easy to overcome.

Fortunately, the learnings that have come about from past experiences provide us with great insights that help in addressing such challenges.

In this book, we have outlined a governance framework that, if followed, will help you steer the journey successfully. This framework is based on real-life learnings from the authors' experiences and published cases.

Following is a summary of the components of this governance framework with the key takeaway from each component:

1. **Strategy and execution:** Have a clear purpose, vision, strategy, and detailed execution plan. This is the starting point that you must get right. Failure to do so will lead to a high degree of risk and failure probability. We have outlined earlier in the book in detail how to do this successfully.

2. **Supporting ecosystem:** One of the key requirements for a successful journey is to gain the support of everyone in the ecosystem. This includes stakeholders, customers and partners, employees, technology providers, and authorities.

   Making use of all the support available is necessary as many challenges can only be overcome with the support of those in the ecosystem.

3. **Change management:** Many of the failures have been contributed to the failure in managing the change that digital transformation brings about.

Special attention must be given to change management by defining a strategy from day one. This strategy must include the impact that digital transformation will have on each party in the ecosystem. One main party that is often forgotten is the customer.

Change management is a journey on its own.

4. **Frameworks, standards, and the 3 Ps:** Standards and frameworks provide you with a proven methodology to adapt and rely on. They also provide you with a reference point for comparison, benchmarking, and audit at any time. They prove to be essential for world-class implementation.

    The 3 Ps (Processes, Policies, and Procedures) are the pillars upon which the frameworks and standards are based. They are the next-level operational aspects that ensure adherence to the standards and frameworks. Defining those Ps right will ensure a smooth ride through the journey.

5. **Performance management and control:** Monitoring execution goes without saying. However, when it comes to digital transformation, this also becomes a major challenge as there are many parameters to watch and it can become confusing.

Therefore, we have outlined how to identify the key success factors and monitor them. Those are the ones that will have a major impact on the outcome. Watching what matters leads to making the right decisions in a timely matter.

6. **Cyber Security:** The digital world is not always safe. Doing business digitally poses many risks and you need to protect your organization and more importantly your customers' information.

   Some laws and regulations detail what you must protect. The challenge is that those regulations defer from country to country, and you need to be aware of all of them.

   Some standards help in defining an information security management system. Those standards provide great guidelines and audit references that can be used to assess the maturity of the security system.

   Most organizations overspend when it comes to cyber security and buy far too many technologies and solutions. We have outlined how to optimize and balance your risk/return investment in cyber security.

7. **Emerging Technologies:** By the time you have your first digital business transaction, you will find that new

technologies have come. This has been a great challenge for many organizations and has caused many turbulences during execution.

A good technology strategy will address this issue. Knowing which technology to use and when is important so you do not end up as a technology shop. It is important to remember that new technologies need time to mature and that not all new technologies live long.

We have provided guidelines for using technologies and touched upon some of the emerging technologies.

8. **Talent Management:** There is no doubt that without having the right talent throughout the digital transformation journey, success will be difficult to achieve. Everybody is competing to get and retain talent to stay ahead in the game.

   Contrary to the belief that the latest technology is the key to competitive advantage in the digital world, the reality is that talent is what makes the real difference. This is not to say that technology is not important, but technology without the right talent is like having the best book and keeping it on the shelf due to an inability to read it.

   An important point is that many types of skills are needed, not just technology ones. Areas such as data management, data analysis, process design, and others are very important and need to be acquired and retained.

Remember, no technology can replace talent. To succeed, you need to define and implement a strategy to ensure that you have the essential skills that you need.

9. **Data Governance:** Data is the oil of the digital landscape. The ability to analyze data and use it to create value starts from data design, definition, acquisition, and usage.

To be able to use data, it must be trusted, available, secure, and understood. Therefore, it is important to ensure that the data of the organization meet the above properties.

Misuse of data poses a lot of risk to the organization. Therefore, there are many laws and regulations such as privacy laws or GDPR (General Data Protection Regulation) that protect against the misuse of data pertaining to customers, employees, vendors, shareholders, or any other stakeholder.

All the above can be addressed by implementing a data governance framework that outlines policies, policies, and guidelines for how data is used.

GLOSSARY

ADKAR:

This is a change management methodology created by Jeffrey Hiatt, the founder of Prosci. It defines five stages known as ADKAR (Awareness, Desire, Knowledge, Ability, and Reinforcement)

Change Management:

refers to a set of systems and processes that enable the organization and all the ecosystem layers make the transition from current state to the newly desired state effectively and efficiently

CSOC:

Cyber Security Operations Centre – a centralized facility from where the computer infrastructure and software is monitored in cybersecurity and alert given should there be any threat

Cyber Security:

Deals with protecting organization assets and looks at ensuring availability of systems and data. In addition, it protects data from unauthorized access to ensure integrity and confidentiality

Data Governance:

Consists of policies, procedures, and guidelines that control how data is managed to ensure that data is consistent and trustworthy and does not get misused

Digital Governance:

Refers to a set of measures put in place to ensure the successful execution of digital initiatives

Digitization:

This is the use of digital technologies and data to improve existing business processes and create a digital culture where digital data is at the core of the process

Digitizing:

This is the process of converting information from a physical format into a digital one

Digital transformation:

This is a journey towards digital business. It involves a redefinition of the business operating model. Digital transformation includes both digitization and digitalization, using these two processes to transform the way that an enterprise conducts its business

Digital Strategy:

Outlines how the business is going to run digitally, and how it is expected to achieve the organization's vision, overall business objectives, and goals

Ecosystem:

is a purpose-based, seamlessly connected, and interrelated system of entities working for a common purpose

Framework:

A structured approach towards addressing your needs and achieving the desired outcomes, and is used to provide guidelines and best practices

Industry 4.0:

also referred to as the fourth industrial revolution is more relevant in the manufacturing segment and deals with machines interconnectivity with digital systems

KPI:

A quantifiable measurement of an important parameter that is a 'must have' for successfully achieving a target or goal.

KRA:

Key Results Area – activities you must do to fulfill your responsibilities and achieve business goals

Performance Management:

A set of policies, guidelines, and monitoring systems that ensure that individuals, teams, and the overall organization perform according to the defined goals and objectives

Project Charter:

A project charter is the document that contains all strategic and tactical details of any project and is jointly agreed upon between all key involved parties of the project and outlines the scope and framework of the project

RFP:

Stands for Request for Proposal. This document contains all strategic information about the IT project or Digital Transformation project and is endorsed by the CEO, project leadership, and consulting leadership team.

Standard:

This is a model or guideline towards a topic, product, process, or procedure that has gained global acceptance and has proven its usefulness

Supporting ecosystem:

a logical group of all such components, which are needed to execute the digital transformation project. The components are the technology professionals, employees, vendors, technology itself, hardware and computing environment

Super Users:

all such employees, who possess good business knowledge and are dedicated to IT and digital transformation programs to provide the business inputs to consultants for system design and configuration. Such employees are dedicated full-time to digital projects in most cases.

REFERENCES

"Steps in Building a Digital Transformation Strategy to Grow Your Business | Finance Magnates" https://www.financemagnates.com/thought-leadership/steps-in-building-a-digital-transformation-strategy-to-grow-your-business/amp/

"Digital transformation: 4 strategy questions to ask | The Enterprisers Project" https://enterprisersproject.com/article/2021/3/digital-transformation-4-strategy-questions

"The key pillars of a digital transformation strategy | IT PRO" https://www.itpro.co.uk/business-strategy/digital-transformation/359736/the-key-pillars-of-a-digital-transformation?amp

"How Digital Transformation is Driving The Customer Experience" https://www.superoffice.com/blog/digital-transformation/

"Spending on digital transformation technologies and services worldwide from 2017 to 2025 (in trillion U.S. dollars)" https://www.statista.com/statistics/870924/worldwide-digital-transformation-market-size/

"Digital Transformation Change Management: How to Build On Initial Success | PLANERGY Software" https://planergy.com/blog/digital-transformation-change-management/

"5 Biggest Challenges to a Successful Digital Transformation |" https://www.netsolutions.com/insights/challenges-to-a-successful-digital-transformation-and-how-to-overcome-them/

"70% of Digital Transformations Fail, and Here's Why – Consulteer" https://www.consulteer.com/blog/2020/09/16/70-of-digital-transformations-fail-and-heres-why/

"Appinventiv's Guide to Shaping Your Digital Transformation Strategy" https://appinventiv.com/blog/appinventiv-digital-transformation-guide/amp/

"5 Stories That Demonstrate the Power of Digital Transformation | Azeus Convene" https://www.azeusconvene.com/articles/five-success-stories-that-demonstrates-the-power-of-digital-transformation

"10 digital transformation challenges and how to overcome them | IDG Connect" https://www.idgconnect.com/article/3610398/10-digital-transformation-challenges-and-how-to-overcome-them.amp.html

"3 Key Case Studies for Successful Digital Transformation" https://blog.remesh.ai/3-key-case-studies-for-successful-digital-transformation?hs_amp=true

"How to overcome Digital transformation challenges | Blog | CodeCoda" https://codecoda.com/en/blog/entry/how-to-overcome-digital-transformation-challenges

"Five reasons why digital transformation efforts fail – The Financial Express" https://www.financialexpress.com/brandwagon/five-reasons-why-digital-transformation-efforts-fail/2221109/lite/

"4 Case Studies in Digital Transformation Strategy – RPA | UiPath" https://www.uipath.com/blog/digital-transformation/digital-transformation-strategy-case-studies

"Top 10 Digital Transformation Failures of All Time, Selected by an ERP Expert Witness – Third Stage Consulting Group" https://www.thirdstage-consulting.com/top-10-digital-transformation-failures-of-all-time-selected-by-an-erp-expert-witness/

"Top 5 Reasons Digital Transformation Efforts Fail | Pandio" https://pandio.com/blog/top-5-reasons-digital-transformation-efforts-fail/

"The case of digital transformation success and failure" https://www.cigniti.com/blog/digital-transformation-strategy-success-failure/amp/

"7 Digital Transformation Challenges & How To Overcome Them" https://www.panorama-consulting.com/digital-transformation-challenges/

"5 Change Management Strategies for Digital Transformation" https://www.impactmybiz.com/blog/blog-5-change-management-strategies-for-digital-transformation/

"The 5 digital transformation pillars – Headspring" https://headspring.com/2020/12/10/the-5-pillars-of-digital-transformation/

"How to structure an effective digital transformation plan | Smart Insights" https://www.smartinsights.com/manage-digital-transformation/digital-transformation-strategy/structure-effective-digital-transformation-plan/amp/

"If you're not building an ecosystem, chances are your competitors are" https://www.mckinsey.com/business-functions/strategy-and-corporate-finance/our-insights/the-strategy-and-corporate-finance-blog/if-youre-not-building-an-ecosystem-chances-are-your-competitors-are

"How To Drive Digital Transformation Through Organizational Change Management (Ocm) And Itil 4" https://www.beyond20.com/blog/digital-transformation-with-organizational-change-management-and-itil-4/

"Culture is key to digital transformation success" https://blog.softwareag.com/culture-digital-transformation

"Digital Transformation Comes Down to Talent in 4 Key Areas" https://hbr.org/2020/05/digital-transformation-comes-down-to-talent-in-4-key-areas

"Digital transformation is about People & Talent" https://www.cio.com/article/189377/digital-transformation-is-about-people-and-talent.html

"Build a talent strategy for digital transformation" https://www.kornferry.com/insights/featured-topics/resources/build-a-talent-strategy-for-digital-transformation

"How digital transformation is driving a shift in talent and culture" https://www.ey.com/en_us/cio/how-digital-transformation-is-driving-talent-and-culture

"Disrupt or be disrupted: The right talent for digital transformation" https://www.pluralsight.com/blog/career/digital-dna

"Digital Transformation Is About Talent, Not Technology" https://hbr.org/2020/05/digital-transformation-is-about-talent-not-technology

"Data Is Essential To Digital transformation" https://www.forbes.com/sites/forbestechcouncil/2020/12/03/data-is-essential-to-digital-transformation/?sh=20e634e126c9

"What is data governance and why does it matter?" https://searchdatamanagement.techtarget.com/definition/data-governance

"Real digital transformation starts with data governance" https://www.itproportal.com/features/real-digital-transformation-starts-with-data-governance/

"Data Governance Enables Digital Transformation" https://www.linkedin.com/pulse/data-governance-enables-digital-transformation-ali-cpa-cia-cissp/

"What role will Data Governance play in the Digital Transformation after COVID-19?" https://business.blogthinkbig.com/what-role-will-data-governance-play-in-the-digital-transformation-after-covid-19/

"What Is Data Governance? Why Do I Need It?" https://www.bmc.com/blogs/data-governance/

"How do companies create value from digital ecosystems?" https://www.mckinsey.com/business-functions/mckinsey-digital/our-insights/how-do-companies-create-value-from-digital-ecosystems

ACKNOWLEDGMENTS – AHMAD

I would like to thank many people who have been part of lifelong learning that made writing this book possible.

I would like to thank my wife and my children who have always supported me throughout this journey.

Special thanks to Emirates Global Aluminum (EGA), the organization that has provided me with endless opportunities of learning throughout my career. Starting from the CEOs, top leadership team, and my great IT team.

Dr. Saeed Al Dhaheri (chairman) and Abdulqader Ali (CEO) of Smartworld who made it possible to establish CIOMajlis (A platform whereby CIOs in the UAE meet and network).

The many fellow CIOs that I have interacted through exchanging visits, conferences, and events.

My co-author Arun Tewary, with whom I had many brainstorming and debate sessions.

Special thanks to Passionpreneur Publishing: specifically Moustafa Hamwi, the CEO for his inspirational encouraging words and support that provided us with energy, and Clare Mclvor for great coaching and guiding sessions.

ACKNOWLEDGMENTS – ARUN

Thanks to my wife, who kept doing the progress check on this book journey and kept reminding me. To my children who continued to encourage me.

Special thanks to my father-in-law, who was an academician and kept guiding me and to my mother-in-law who kept inspiring me.

Thanks to all my industry colleagues from whom I kept learning,

Thanks to my seniors throughout my professional journey, who were always there with me to keep sharpening my skills. Special mentions to Akhil Pandey, Tata Steel, AP Kastuar, Tata Steel, John Helmand, Al Futtaim, AJ Jaganathan Emaar, Saeed Mohammad, Emirates Flight Catering, Salivati from Emirates Flight catering, Sunil Paul from Finesse and Lalu Mahtani from Alpine Creations.

Sincere thanks to my co-author Ahmad, who displayed immense patience during the course of book writing, with great synergy.

And special thanks to Passionpreneur Publishing: to Moustafa Hamwi, the CEO for his inspirational encouraging words and support that provided us with energy, and Clare Mclvor for great coaching and guiding sessions.

EXTRAS

Ahmad Almulla

Contact Details:
Email: amulla_ahmad@yahoo.com
LinkedIn: https://www.linkedin.com/in/almullaahmad/?original
Subdomain=ae
Mobile: +971 50 50 9100

Arun Tewary

Contact Details:
Place – Dubai, UAE
Mobile +97150-4578723
Email arunktewary@hotmail.com
LinkedIn www.linkedin.com/in/aruntewary
Twitter @arunktewary

ABOUT THE AUTHORS

Ahmad Almulla and Arun Tewary, are two Information Technology professionals who are experienced industry veterans from the Middle East. They carry a combined experience of over six decades.

Over the years, they have experienced numerous successes and failures as their careers traversed the most profound period of technological change in history.

Ahmad was the Executive Vice President of Information Technology at Emirates Global Aluminium (One of the largest Aluminium companies in the world), and Arun was the Vice President of Information Technology in Emirates Flight Catering (Part Of Emirates Group). Now, having said goodbye to our previous jobs, they could not contain the urge to share their experience with fellow professionals, CEOs, and executives of the corporate world. They have put genuine efforts to provide you a guideline on your digital transformation journey extracted from our learning over the last three decades or so.

Ahmad Almulla

Ahmad Almulla is a known authority in the digital world. With over 30 years of experience in IT and business coupled with international exposure, Ahmad is considered one of the leading figures when it comes to innovation and creative thinking.

He is one of few who have combined experience in Information Technology and Supply Chain Management, coupled with strong leadership capabilities. This combination and exposure have enabled Ahmad to play a crucial strategic role in the success of Dubai Aluminum (DUBAL) and Emirates Global Aluminum (EGA).

Ahmad is credited as one of the driving forces behind the IT revolution at EGA, and a visionary leader who has ushered in an era of technological change in the company. Over 30 years of service at EGA, Ahmad has been at the helm of various strategic initiatives envisaged to position EGA as a leader in technology and information-based infrastructure, and supply chain management. His experience and knowledge in Cyber Security domain has shielded EGA domain and information assets.

During his career, Ahmad has been a frequent keynote speaker at many conferences and symposiums globally, and has led many initiatives that have resulted in EGA receiving several prestigious awards locally, regionally, and internationally.

He has been a pivotal member of the Executive Management Team at EGA for over 15 years, at which level he has contributed strategic expertise in global information systems, information system architecture, supply chain strategy, and automation procedures.

Drawing on his insight into technology management and ability to create highly effective IT frameworks, Ahmad has effectively leveraged IT as a key catalyst for EGA's exponential growth over the past two decades. This has been achieved through optimizing operational proficiency via innovative use of technology.

Ahmad also led the vision for a transformation program through the implementation of SAP in DUBAL/EGA and later the integration of multiple SAP systems. The latter system not only revolutionized the way DUBAL/EGA business departments function by using world-class best practices but also positioned them for future growth.

As part of community support, Ahmad has played a crucial role in establishing CIOMajlis (A platform for CIOs to network and discuss IT-related topics) and was the founding chairman. He also served as the vice-chairman of Dubai Quality Group (DQG). In addition, he is a board member of the Association of Change Management Professionals (ACME) Middle East Chapter.

Ahmad holds an MBA from the University of New England and a bachelor's degree in Computer Engineering from the University of Arizona. He has completed the PED (Programme for Executive Development) at IMD Business School in Switzerland. He completed the Directors Development Programme at Mudara Institute of Directors. He has also been a frequent nominee for international training sessions at prestigious institutes such as the IMD in Lausanne, Switzerland.

Currently, Ahmad advises Boards of Directors, CEOs, and CIOs on IT strategy, cyber security, IT governance, and IT-led innovation.

He serves as a board member of Paramount Computer Systems. He is a strategic advisor to HCL Middle Ease. He also is a board member of CIOMajlis and a member of the IDC Middle East advisory council.

Arun Tewary

Arun Tewary is a well-known name in the IT industry in the Middle East, where he has been active for the last 25 years. But he did not limit his interactions and exposure only to the Middle East; he has also remained in active communications beyond geographies.

He is a performance-driven professional with proven success in leading, managing, and overseeing technological transformation globally, as well as enabling worldwide business transformational milestones.

Arun is a Mechanical Engineer through education and is one of the early adopters in the profession of Information Technology. He says, "when I was knocking at the entrance of Information Technology profession, Bill Gates with Paul Allen were busy setting up Microsoft". The technology was heavily influenced by engineering culture, which set up the industry for a journey of rapid growth. People also call it the Fourth Information revolution. Arun has lived through this revolution from its early days, covering the full range from Mainframe to Cloud, from heavy codes to no / low codes, from relational databases, ISAM files to No SQL, and object-oriented databases.

Arun is experienced in creating new business lines, facing digital transformation challenges, executing new IT operating models, and implementing technology plans. He is well-known for strategy planning and execution and advisory/consulting service with a stellar record of serving as a Board Member for multiple groups as well as councils. He also has demonstrated the ability to create global digital business platforms, ensuring organizational transformation, implementing intelligent hybrid models, and deploying SAP, JD Edwards, Oracle ERP, and other ERP systems. He ensured the largest SAP implementation in the Middle East.

Furthermore, he is skillful in cultivating cordial relationships with key stakeholders to ensure continuous growth and success. He is instrumental in communicating with excellent interpersonal and analytical skills. He is adept at managing teams of professionals internationally while ensuring professional growth.

Throughout his career, he has attained comprehensive IT transformation experience that adds depth and practical value to the content of this book.

Arun has been a very active contributor to the world of Information Technology through his regular contributions in journals, participation in professional events, and being an active speaker at various forums. He has been recognized by the industry through multiple awards like "Winner of Aviation Technology Awards" in 2012, and the "Arab Technology Award" in 2011 for best implementation in Travel & Hospitality sector.

Arun has been a speaker in various forums like Dell Next, Oracle Open World San Francisco, IDC CIO Round Table panel discussion, IDC IT managers' forum, Middle East CIO Summit by IDC, ACN CIO summit panel discussion, and many more.

Arun started his career as a Mechanical Engineer on the shop floor of a paper manufacturing company in India. Then he moved to a large government-owned power generation company. He joined as a computer programmer in a Steel manufacturing company, which was the start of a professionally satisfying and illustrious career. After 40 years, Arun is still going strong in the domain of Information Technology. Starting from Cobol

Programmer, Arun rose to various positions such as Systems Analysts, Project Manager, Head of IT, and then CIO in large organizations internationally. Arun has consistently worked for very large and reputed organizations. Presently, he is Strategy Advisor and director for Finesses Global and Director-IT with Alpine Creations. He is also independent advisor to various companies and on the board of a few companies.

Having acquired so much professional experience through his fulfilling professional journey, Arun decided to share his experience with the industry – which is exactly where Arun and Ahmad joined hands in this journey.

www.ingramcontent.com/pod-product-compliance
Lightning Source LLC
Chambersburg PA
CBHW060931220326
41597CB00020BA/3474

9781761240461